T0353343

Emerging Trends and Applications in Cognitive Computing

Pradeep Kumar Mallick
St. Peter's University, India

Samarjeet Borah
Sikkim Manipal University, India

A volume in the Advances in
Computational Intelligence and
Robotics (ACIR) Book Series

Published in the United States of America by
        IGI Global
        Engineering Science Reference (an imprint of IGI Global)
        701 E. Chocolate Avenue
        Hershey PA, USA 17033
        Tel: 717-533-8845
        Fax: 717-533-8661
        E-mail: cust@igi-global.com
        Web site: http://www.igi-global.com

Library of Congress Cataloging-in-Publication Data

Names: Mallick, Pradeep Kumar, 1984- editor. | Borah, Samarjeet, editor.
Title: Emerging trends and applications in cognitive computing / Pradeep
  Kumar Mallick and Samarjeet Borah, editors.
Description: Hershey, PA : Engineering Science Reference, [2019]
Identifiers: LCCN 2017055229| ISBN 9781522557937 (h/c) | ISBN 9781522557944
  (eISBN)
Subjects: LCSH: Artificial intelligence--Handbooks, manuals, etc. |
  Computational intelligence--Handbooks, manuals, etc. | Biomedical
  engineering--Data processing--Handbooks, manuals, etc. | Neural
  computers--Handbooks, manuals, etc. | Cognitive science--Handbooks,
  manuals, etc.
Classification: LCC TA347.A78 H36 2019 | DDC 006.3--dc23 LC record available at https://lccn.
loc.gov/2017055229

This book is published in the IGI Global book series Advances in Computational Intelligence and Robotics (ACIR) (ISSN: 2327-0411; eISSN: 2327-042X)

British Cataloguing in Publication Data
A Cataloguing in Publication record for this book is available from the British Library.

All work contributed to this book is new, previously-unpublished material.
The views expressed in this book are those of the authors, but not necessarily of the publisher.

For electronic access to this publication, please contact: eresources@igi-global.com.

Advances in Computational Intelligence and Robotics (ACIR) Book Series

ISSN:2327-0411
EISSN:2327-042X

Editor-in-Chief: Ivan Giannoccaro, University of Salento, Italy

MISSION

While intelligence is traditionally a term applied to humans and human cognition, technology has progressed in such a way to allow for the development of intelligent systems able to simulate many human traits. With this new era of simulated and artificial intelligence, much research is needed in order to continue to advance the field and also to evaluate the ethical and societal concerns of the existence of artificial life and machine learning.

The **Advances in Computational Intelligence and Robotics (ACIR) Book Series** encourages scholarly discourse on all topics pertaining to evolutionary computing, artificial life, computational intelligence, machine learning, and robotics. ACIR presents the latest research being conducted on diverse topics in intelligence technologies with the goal of advancing knowledge and applications in this rapidly evolving field.

COVERAGE

- Computational Logic
- Pattern Recognition
- Intelligent control
- Evolutionary computing
- Automated Reasoning
- Machine Learning
- Agent technologies
- Computational Intelligence
- Cyborgs
- Adaptive and Complex Systems

IGI Global is currently accepting manuscripts for publication within this series. To submit a proposal for a volume in this series, please contact our Acquisition Editors at Acquisitions@igi-global.com or visit: http://www.igi-global.com/publish/.

Titles in this Series

For a list of additional titles in this series, please visit:
https://www.igi-global.com/book-series/advances-computational-intelligence-robotics/73674

Nature-Inspired Algorithms for Big Data Frameworks
Hema Banati (Dyal Singh College, India) Shikha Mehta (Jaypee Institute of Information Technology, India) and Parmeet Kaur (Jaypee Institute of Information Technology, India)
Engineering Science Reference • ©2019 • 412pp • H/C (ISBN: 9781522558521) • US $225.00

Novel Design and Applications of Robotics Technologies
Dan Zhang (York University, Canada) and Bin Wei (York University, Canada)
Engineering Science Reference • ©2019 • 341pp • H/C (ISBN: 9781522552765) • US $205.00

Optoelectronics in Machine Vision-Based Theories and Applications
Moises Rivas-Lopez (Universidad Autónoma de Baja California, Mexico) Oleg Sergiyenko (Universidad Autónoma de Baja California, Mexico) Wendy Flores-Fuentes (Universidad Autónoma de Baja California, Mexico) and Julio Cesar Rodríguez-Quiñonez (Universidad Autónoma de Baja California, Mexico)
Engineering Science Reference • ©2019 • 433pp • H/C (ISBN: 9781522557517) • US $225.00

Handbook of Research on Predictive Modeling and Optimization Methods in Science...
Dookie Kim (Kunsan National University, South Korea) Sanjiban Sekhar Roy (VIT University, India) Tim Länsivaara (Tampere University of Technology, Finland) Ravinesh Deo (University of Southern Queensland, Australia) and Pijush Samui (National Institute of Technology Patna, India)
Engineering Science Reference • ©2018 • 618pp • H/C (ISBN: 9781522547662) • US $355.00

Handbook of Research on Investigations in Artificial Life Research and Development
Maki Habib (The American University in Cairo, Egypt)
Engineering Science Reference • ©2018 • 501pp • H/C (ISBN: 9781522553960) • US $265.00

Critical Developments and Applications of Swarm Intelligence
Yuhui Shi (Southern University of Science and Technology, China)
Engineering Science Reference • ©2018 • 478pp • H/C (ISBN: 9781522551348) • US $235.00

For an entire list of titles in this series, please visit:
https://www.igi-global.com/book-series/advances-computational-intelligence-robotics/73674

701 East Chocolate Avenue, Hershey, PA 17033, USA
Tel: 717-533-8845 x100 • Fax: 717-533-8661
E-Mail: cust@igi-global.com • www.igi-global.com

Table of Contents

Detailed Table of Contents

Chapter 1

This chapter focuses on clinical decision system (CDS) uses in healthcare units. In this chapter, cognitive approaches are taken using soft computing techniques to design clinical decision systems (CDS) for modern healthcare units. Cognitive computing-based approach is considered. It focuses on cardiac disease detection exclusively by considering its surrounding factors. Fuzzy logic is utilized as one part. The other part includes diabetic detection using deep neural network (DNN) for the automatic identification of the disease. The experiment was done with the Pima Indian dataset. The classification result has been presented in the result section. The decision system in the healthcare unit is a suitable example of a multi-agent system.

Chapter 2

The focus of this chapter is to design a cognitive information retrieval (CIR) framework using inference engine (IE). IE permits one to analyze the central concepts of information retrieval: information, information needs, and relevance. The aim is to propose an inference engine in which adequate user preferences are considered. As the cognitive inference engine (CIE) approach is involved, the complex inquiries are required to return more important outcomes as opposed to customary database questions which get irrelevant and unsolicited responses or results. The chapter highlights the framework of a cognitive rule-based engine in which preference queries are dealt with while keeping in mind the intention of the user, their performance, and optimization.

Chapter 3
    Heath Yates, Kansas State University, USA
    Brent Chamberlain, Utah State University, USA
    William Baldwin, Biosecurity Research Institute, USA
    William Hsu, Kansas State University, USA
    Dana Vanlandingham, Biosecurity Research Institute, USA

Affective computing is a very active and young field. It is driven by several promising areas that could benefit from affective intelligence such as virtual reality, smart surveillance, perceptual interfaces, and health. This chapter suggests new design for the detection of animal affect and emotion under an affective computing framework via mobile sensors and machine learning. The authors review existing literature and suggest new use cases by conceptual reevaluation of existing work done in affective computing and animal sensors.

Chapter 4
    Sushruta Mishra, KIIT University, India
    Soumya Sahoo, C. V. Raman College of Engineering, India
    Brojo Kishore Mishra, C. V. Raman College of Engineering, India

The modern techniques of artificial intelligence have found application in almost all the fields of human knowledge. Among them, two important techniques of artificial intelligence, fuzzy systems (FS) and artificial neural networks (ANNs), have found many applications in various fields such as production, control systems, diagnostic, supervision, etc. They evolved and improved throughout the years to adapt arising needs and technological advancements. However, a great emphasis is given in the engineering field. The techniques of artificial intelligence based on fuzzy logic and neural networks are frequently applied together for solving engineering problems where the classic techniques do not supply an easy and accurate solution. Separately, each one of these techniques possesses advantages and disadvantages that, when mixed together, provide better results than the ones achieved with the use of each isolated technique. As ANNs and fuzzy systems have often been applied together, the concept of a fusion between them started to take shape. Neuro-fuzzy systems were born which utilize the advantages of both techniques. Such systems show two distinct ways of behavior. In a first phase, called learning phase, it behaves like neural networks that learn internal parameters off-line. Later, in the execution phase, it behaves like a fuzzy logic system. A neuro-fuzzy system is a fuzzy system that uses a learning algorithm derived from or inspired by neural network theory to determine its parameters (fuzzy sets and fuzzy rules) by processing data samples. Neural networks and fuzzy systems can be combined to join its advantages and to cure its individual illness. Neural networks introduce its computational characteristics of learning in the fuzzy systems and receive from them the interpretation and clarity of systems representation. Thus, the disadvantages of the fuzzy systems are compensated by the capacities of the neural networks. These techniques are complementary, which justifies its use together. This chapter deals with an analysis of neuro-fuzzy systems. Benefits of these systems are studied with its limitations too. Comparative analyses of various categories of neuro-fuzzy systems are discussed in detail. Apart from these, real-time applications of such systems are also presented.

Chapter 5

This chapter presents an approach to combine probability distributions with imprecise (fuzzy numbers) parameters (mean and standard deviation) as well as fuzzy numbers (FNs) of various types and shapes within the same framework. The amalgamation of probability distribution and fuzzy numbers are done by generating three algorithms. Human health risk assessment is performed through the proposed algorithms. It is found that the chapter provides an exertion to perform human health risk assessment in a specific manner that has more efficacies because of its capacity to exemplify uncertainties of risk assessment model in its own fashion. It affords assistance to scientists, environmentalists, and experts to perform human health risk assessment providing better efficiency to the output.

Chapter 6

With the increase in the number of computers, the amount of energy consumed by them is on a significant rise, which in turn is increasing carbon content in atmosphere. With the realization of this problem, measures are being taken to minimize the power usage of computers. The solution is green computing. It is the efficient utilization of computing resources while minimizing environmental impact and ensuring both economic and social benefits. Green computing is a balanced and sustainable approach towards achieving a healthier and safer environment without compromising the technological needs of the current and future generations. This chapter studies the architectural aspects, the scope, and the applications of green computing. The emphasis of this study is on current trends in green computing, challenges in the field of green computing, and the future trends of green computing.

Chapter 7

*Dalila Amara, SMART Lab, Université de Tunis, Institut Supérieur de
Gestion, Tunis, Tunisie*
*Latifa Ben Arfa Rabai, SMART Lab, Université de Tunis, Institut
Supérieur de Gestion, Tunis, Tunisie & College of Business,
University of Buraimi, Al Buraimi, Oman*

Software measurement helps to quantify the quality and the effectiveness of software to find areas of improvement and to provide information needed to make appropriate decisions. In the recent studies, software metrics are widely used for quality assessment. These metrics are divided into two categories: syntactic and semantic. A literature review shows that syntactic ones are widely discussed and are generally used to measure software internal attributes like complexity. It also shows a lack of studies that focus on measuring external attributes like using internal ones. This chapter presents a thorough analysis of most quality measurement concepts. Moreover, it makes a comparative study of object-oriented syntactic metrics to identify their effectiveness for quality assessment and in which phase of the development process these metrics may be used. As reliability is an external attribute, it cannot be measured directly. In this chapter, the authors discuss how reliability can be measured using its correlation with syntactic metrics.

Chapter 8

Smruti Rekha Das, SOA University, India
Kuhoo, College of Engineering and Technology, India
Debahuti Mishra, SOA University, India
Pradeep Kumar Mallick, Gurukula Kangri Vishwavidyalaya, India

The basic aim of risk management is to recognize, assess, and prioritize risk in order to assure that the uncertainty should not deviate from the intended purpose of the business goals. Risk can take place from various sources, which includes uncertainty in financial markets, recessions, inflation, interest rates, currency fluctuations, etc. Various methods used for this management of risk are faced with various decisions such as the market price, historical data, statistical methodologies, etc. For stock prices, the information derives from the historical data where the next price depends only upon the current price and some of the outside factors. Financial market is very risky to invest money, but the proper prediction with handling the risk will benefit a lot. Various types of risk in the financial market and the appropriate solutions to overcome the risk are analyzed in this study.

Chapter 9

Clustering is a process of grouping a set of data points in such a way that data points in the same group (called cluster) are more similar to each other than to data points lying in other groups (clusters). Clustering is a main task of exploratory data mining, and it has been widely used in many areas such as pattern recognition, image analysis, machine learning, bioinformatics, information retrieval, and so on. Clusters are always identified by similarity measures. These similarity measures include intensity, distance, and connectivity. Based on the applications of the data, different similarity measures may be chosen. The purpose of this chapter is to produce an overview of much (certainly not all) of clustering algorithms. The chapter covers valuable surveys, the types of clusters, and methods used for constructing the clusters.

Chapter 10

In the IoTs era, the short-range mobile transceivers will be implanted in a variety of daily requirements. In this chapter, a detail survey in several security and privacy concerns related to internet of things (IoTs) by defining some open challenges are discussed. The privacy and security implications of such an evolution should be carefully considered to the promising technology. The protection of data and privacy of users has been identified as one of the key challenges in the IoT. In this chapter, the authors present internet of things with architecture and design goals. They survey security and privacy concerns at different layers in IoTs. In addition, they identify several open issues related to the security and privacy that need to be addressed by research community to make a secure and trusted platform for the delivery of future internet of things. The authors also discuss applications of IoTs in real life. A novel approach based on cognitive IoT is presented, and a detailed study is undertaken. In the future, research on the IoTs will remain a hot issue.

Foreword

Cognitive Informatics (CI) is a burgeoning interdisciplinary domain comprising of the cognitive and information sciences that focuses on human information processing, mechanisms and processes within the context of computing and computer applications.

The main goal is to research and develop technologies to facilitate and extend the information management capacity of individuals through the development and application of novel concepts in human-system integration to address cognitive bottlenecks (e.g., limitations in attention, memory, learning, comprehension, visualization abilities, and decision making). Such mitigations may include applications and technologies informed by research in psychology/behavioural science, neuroscience, artificial intelligence, or linguistics.

As initiated by Dr. Pradeep Kumar Mallick and Dr. Samarjeet Borah, Handbook of Research on Emerging Trends and Applications in Cognitive Computing presents new approaches and methods for solving real-world problems. It offers, in particular, exploratory research that describes novel approaches in the fields of Cognitive Informatics, Cognitive Computing, Computational Intelligence, Advanced Computing, Hybrid Intelligent Models and Applications. New algorithms and methods in a variety of fields are also presented, together with solution-based approaches. The topics addressed include various theoretical aspects and applications of Computer Science, Artificial Intelligence, Cybernetics, Automation Control Theory and Software Engineering.

People are working on these with enthusiasm, tenacity, and dedication to develop new methods of analysis and provide new solutions to keep up with the ever-changing cognitive informatics. This book is a good step in Cognitive Computing direction.

Gyoo-Soo Chae
Baekseok University, South Korea

Preface

Emerging Trends and Applications in Cognitive Computing is a collection of research findings of various authors working in the domain.

The first chapter of this volume discusses a design of cognitive healthcare system for coronary cardiac diseases detection. It is basically a Cognitive Information Retrieval (CIR) framework using Inference Engine (IE). It explains a framework of a cognitive rule-based engine in which preference queries are dealt with keeping in mind the intention of the user, their performance and optimization. Next chapter is cognitive radio network spectrum ("white spaces") holes sensing, assignment methods and access in dual technologies. As per the authors prime challenge for optional radio frameworks is to have the capacity to robustly detect when they are inside such a frequencies hole. We review existing literature and suggest new use cases by conceptual reevaluation of existing work done in affective computing and animal sensors. A study on neuro-fuzzy models and applications is presented by Mishra et al. in their chapter. Benefits of these systems are studied with its limitations too. They are presenting a comparative analyses of various categories of Neuro-Fuzzy systems are discussed in detail. The evaluation of health risk is the procedure to assess the character and chance of hostile human health consequences those are affected by radiation or other harmful compounds in polluted ecological medium, at the present or in the upcoming days. Palash Dutta discusses the aspects in the chapter human health risk assessment via amalgamation of probability and fuzzy numbers. The nest presents a pattern mining algorithm - FTISPAM using hybrid genetic algorithm. The next chapter studies the architectural aspects, the scope and applications of green computing. A work on software quality management presents a thorough analysis of most quality measurement concepts. Authors also discuss how reliability can be measured using its correlation with syntactic metrics. The basic aim of risk management is to recognize, assess and prioritize risk, in order to assure that the uncertainty should not deviate the intended purpose of the business goals. A study on risk management in financial market is presented by Das et al. The next chapter provides an overview of much of clustering algorithms. The chapter is covering the little much valuable survey, the types of clusters, methods used for constructing the clusters. Mishra et al. are addressing security issues and standards in Internet of Things.

This volume represents a global forum for research on cognitive computing. It includes mostly the current works and research findings from various research labs, universities and institutions and may lead to development of market demanded products. The works reports substantive results on a wide range of learning methods applied to a variety of learning problems. It provides solid support via empirical studies, theoretical analysis, or comparison to psychological phenomena. The volume includes works to show how to apply learning methods to solve important applications problems as well as how machine learning research is conducted.

The volume editors are very thankful to all the authors, contributors, reviewers and the publisher for making this effort a successful one.

Pradeep Kr. Mallick
St. Peter's University, India

Samarjeet Borah
Sikkim Manipal University, India

Chapter 1

Design of Cognitive Healthcare System for Coronary Cardiac Disease Detection

Mihir Narayan Mohanty
Siksha 'O' Anusandhan University, India

ABSTRACT

This chapter focuses on clinical decision system (CDS) uses in healthcare units. In this chapter, cognitive approaches are taken using soft computing techniques to design clinical decision systems (CDS) for modern healthcare units. Cognitive computing-based approach is considered. It focuses on cardiac disease detection exclusively by considering its surrounding factors. Fuzzy logic is utilized as one part. The other part includes diabetic detection using deep neural network (DNN) for the automatic identification of the disease. The experiment was done with the Pima Indian dataset. The classification result has been presented in the result section. The decision system in the healthcare unit is a suitable example of a multi-agent system.

INTRODUCTION

Human being has always been a curious creature, constantly aiming at increasing his knowledge about the world he lives in. Towards achieving new knowledge, the typical process of learning consists thus of following steps: (1) collecting a large amount of observations, (2) extracting relevant information, (3) designing a general model that best explains past and future observations. Machine learning found applications in many fields such as genetics, search engines, natural language processing, computational finance or stock market analysis and computer vision. In

DOI: 10.4018/978-1-5225-5793-7.ch001

the case of medical applications, the interest for learning-based methods seems more recent. Nevertheless, this trend is increasing, and more works containing machine learning as keyword are published each year.

In spite of their effective usefulness, it is not popular clinically that remains a concern to researchers as well as medical practitioners. The objective is to changes in clinical effectiveness related to patient care. The design of the evaluated CDSs relies on either randomized controlled clinical trials (RCTs) or laboratory experiments to determine the performance of the physicians or systems under controlled environment. Lack of elaborate surveys on methodologies that indicates reasons for not using CDSs by clinicians and their practice patterns. It requires the involvement of other medical professionals and computer applications. These are computer based patient records (CPRs), hospital information systems (HISs), ancillary care systems, physician order entry (POE), etc. (Devi, Ramani & Pandian, 2014; Singh, Mohanty & Choudhury, 2013; Mohanty, 2016). Such research seems useful in solving issues related to acceptance or use of CDSs and their relative relevance.

Cognitive informatics (CI) is an area related to cognitive and information sciences. The focus is on processing of human information, computing and computer applications mechanisms as well as processes (Al-Sakran, 2015).In this the CI focuses on understanding of activities and work processes related to human cognition that include the interventional solutions concerned to finding solutions to engineering, computer applications and information technology for better human activities. In the framework of biomedical informatics, CI helps in describing, understanding, and predicting the clinical work activities and its nature that benefits the patients, clinicians as well as lay public. It assists in engineering development and finding computing solutions to boost clinical practice such as efficient decision-support system, patient involvement by providing a tool for timely medication schedule. It also helps public health interventions by providing a suitable mobile application to determine the spread of an epidemic (Ozcift & Gulten, 2011; Chi, Street & Katz, 2010; Su, 2008).

Many attempts have been made to design automatic machines in the field of intelligent and cognitive field. These cognitive machines must know their environments involving similar machines as well as human beings. The machines may vary from self-evidenced practical reasons like computer maintenance expenses to wearable computing in health care which have the cognitive capabilities parallel to the human brain. The aim of this work is to describe the challenges concerned to this new design paradigm which may take into account the systemic problems as well as the design issues. It also includes the teaching of undergraduates in the field of electrical and computer engineering, researchers, etc. Most studies use an experimental or RCT in order to assess system capabilities in a varying clinical environment for better patient care (Silverman et al., 2015). Some of these researches are in the field of CDS whereas none follow a naturalistic design as far as the routine clinical settings concerned to real patients are considered. The studies mainly focus on physicians but not on other clinicians. Further, evaluation of CDS studies is insulated from evaluations of the informatics applications.

Cognitive Technology

Artificial intelligence (AI) technique supports to many technological developments in current scenario of digital world. The technique of cognitive computing has the major role to apply reason in numerous data and by interacting with man and machines. At a decision-making layer, technologies such as machine learning, and deep learning are helping systems interpret information and arrive at effective, informed decisions (Zhang et al., 2018; Liu et al., 2017; Jan et al., 2017).

In healthcare, cognitive applications are helping doctors screen and diagnose patients faster, while allowing companies to broaden their reach across large and dispersed populations.

Machine Learning

Machine learning systems are capable of learning over time without needing to be explicitly programmed. These have capabilities and techniques are helping applications identify patterns in large amounts of information, classify information, make predictions and detect anomalies. These techniques help organizations build systems that can process large amounts of data while applying human-like thinking to information, classify and correlate disparate pieces of information, make more informed decisions and trigger actions in downstream activities.

Deep Learning

Deep learning methods are based on learning representations of data. As a branch of machine learning, deep learning attempts to model high level abstractions in data. It applies multiple layers of processing units where each successive layer uses the output of the previous layer as input. Each layer corresponds to different layers of abstraction, forming a hierarchy of concepts. Deep learning solutions help create applications that can be trained both in a supervised and unsupervised ways. Numerous techniques such as neural networks are under development in the field of deep learning and are expected to be part of current scenario (Craven & Shavlik, 1997)

Neural Networks

Neural networks are among the most used deep learning methods to create learning and reasoning systems. The application of back propagation, as part of neural networks, helps to generate continuous improvement in deep learning algorithms and train multi-layered deep learning systems to enhance their knowledge bases. Neural networks are helping us solve complex problems that typically took skills acquired over many years of learning and human experience. These capabilities are expected to help organizations to analyze audio, video and images as well as any human expert would, only at exponentially faster speeds and at much greater levels of depth and quantity (Al Rahhal et al., 2016; Haykin, 1999; Mohapatra, Palo & Mohanty, 2017).

Cognitive computing represents the state-of-the-art in the management of knowledge and computer-human interaction, integrating a number of advanced disciplines such as artificial intelligence, signal processing, natural language processing, machine learning, speech and vision analysis, and capacities for dialogue and descriptive narrative. The aim of cognitive computing is the development of systems that emulate human behavior, learn for themselves through example and can be deployed in human working environments (Tupe & Kulkarni, 2015; Elayathingal & Sethuramalinga, 2014). The CIis related to multidisciplinary study of cognitive and information sciences. It investigates the mechanisms of human information processing and their engineering applications in the field of computing. The community aims to find technology-based solutions to reduce the workload of a decision maker, increases the throughput and production quality. The computer-based MIS (medical information systems) requires an understanding of computer technology as well as emphasizes on complex behavioral and social processes.

Cognitive science is basically a science research of the fundamental aspects relating to cognition. It involves memory, attention, early language acquisition including applied research. In this context the applied cognitive research helps efficient human–computer interaction (HCI) that benefits the society in many ways including educational research.

In last few decades, have witnessed a few practical and experiential knowledge on the design and implementation that require sensible and intuitive interfaces. It considers a thorough understanding of work direction and consults the potential clinicians to implement in advance.

However, there is a limit to experiential knowledge that can produce the robust generalizations including a sound design as well as implementation principles. Hence, a demand for the common theoretical foundation is essential. The cognitive systems are characterized by two principles that are interdependent: (a) architectural theories that have the potential with respect to every aspects of cognition using unified theory (b) To distinguish different knowledge domains for effectiveness of a given domain.

Functionally, any cognitive system is based on the capabilities it is able to perform such as: (a) focused attention corresponding to the chosen visual characteristics (b) the procedure in which it constrains the performance of human cognitive such as memory limitations (c) The development occurs in its life span. Concerning to the life span problem, a few literature has been focused on cognitive aging as well as its aspects. These are: memory, attention, vision, and change in motor skill as aging functions (Rogers, 2002). It provides potential inputs to informatics related to development of e-health care facilities for seniors, or to patients suffering from chronic health hazards like heart disease, diabetes and arthritis. The field of usability engineering with respect to cognitive task analysis assists to assess cognition corresponding to complex medical environment and computer information systems.

Some features of cognitive systems are adaptiveness, Interactive, Contextual and stateful. The application domains of CI are Multi agent networks, Networks for computational intelligence, Distributed cognitive robots, sensors, and remote control systems (Yan et al., 2008; Nahar et al., 2013). In similar manner models like ANN, FNN, and DNN are able to learn features by means of non-linear data transformations and learn the feature hierarchies. Use of unsupervised training removes over-fitting issues.

Medical Cognition

Advances in technologies related to health information and computing during last two decades brings about many permeate, useful diverse facets in the field of clinical technology. The rapid developments of internet, handheld devices, and wireless technologies can afford new opportunities that support, enhances, and extends user experiences for interactions as well as communications. It is assisted by computer literate health care individuals for better health care. However, the advances in health care domain are slow due to lack of exposure to information technology for use in their working environment. Further, the technologies are hindered by social, cultural, and cognitive issues (Lokanath, Narayan & Srikanta, 2016; AlSharqi et al., 2014; Sarangi, Mohanty & Pattanayak, 2016). There is still a need to adapt the technology in many different diversified levels both in individual and institution category. Absence of such steps may lead to workflow disruptions hence user dissatisfaction.

The knowledge representation, type of search with naturalness proves the success of cognitive computing applications. In healthcare, bio-medical signal and image analysis as well as processing is to increase the portability of screening and diagnosis (Biswal et al., 2016; Biswal et al., 2016). For instance, specialized screening for conditions such as diabetic retinopathy traditionally was limited in availability due to the need for specialized equipment and numerous specialists – thus precluding many patients from being screened preventatively. However, the ability of deep learning systems and neural networks, which learn from previous human diagnosed images, is helping create systems of intelligence that can analyze medical images, predict the risk and existence of such conditions, and recommend either immediate human intervention or offer a reassurance of normality. This would help providers reach larger at-risk populations, many of whom would probably not screen themselves unless the problem was severe (Das, Turkoglu & Sengur, 2009; Ordonez, 2006; Ram & Mohanty, 2017; Hannan et al., 2010).

The pneumonia diagnosis and its severity remains complex due to many interacting factors, thus FCM has been used for decision support. It is suitable for such problems, although, it requires experience and accumulated expert knowledge. It is simple, implementable and requires less time (Sikchi, Sikchi & Ali, 2013; Osowki & Linh, 2001; Pal et al., 2012; Sarangi, Mohanty & Patnaik, 2017). The authors have developed a FCM method

for expert system that can diagnosis infectious diseases, their type and the severity related to cardiac arrest (Sarangi, Mohanty & Patnaik, 2017, 2016; Rahman et al., 2015; Forkan, Khalil & Tari, 2013). The FCM algorithm has been put as our future work based on inputs from clinical data using data mining approaches.

METHODOLOGY

Proposed Multi-Agent System

Patient Agent

Communication gadgets such as a mobile or a personal computer is used to inform the medical authority for hospital registration in case the patient is suffering from a disease. Figure 1 shows the structure of the proposed multi-agent system.

Hospital Registration

A server is used to store the patient parameter during hospital registration and forward these information to a doctor i.e. doctor1 agent who recommends for pathological tests.

Figure 1. Patient consultancy multi-agent structure with a fuzzy system

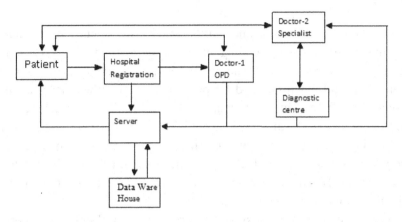

Doctor 1

The doctor 1 agent functions as an OPD of the concerned hospital. With the help of fuzzy IF-THEN rules, this agent suggests for any pathological tests desired of the patient.

Diagnostic Centre Agent

This agent analyses and predicts the disease of the affected person based on the information received from the patient. A two way data flow i.e. both transmitting and receiving is accomplished by this agent.

Doctor 2 or Specialist Agent

This agent communicates using either SMS or e-mail with the concerned doctor and notifies him/her about the abnormality observed in a patient so that the desired prescription is provided. Ultimately, the prescription based on the disease is generated and is delivered to the concerned patient. In case any further investigation is desired by the doctor regarding the patient's medical history or evaluation of patient parameters in a continuous basis, it will made available to the doctor from the server. It helps the doctor for future monitoring of the patient condition.

Server Agent

It provides the desired disease information to all other agents. It communicates with the diagnostic center by sending and receiving patient report. These reports serve as inputs to the physician. The server agent takes the decision on data flow management as and when required for providing efficient healthcare facilities to the patient using disease information and making the data available for the doctor.

The flow of data in the diagnostic agent is shown in Figure 2. As shown in this Figure, the diagnostic center receives information on the patient symptoms. Further analyze and prediction of the disease is made based on the received data in the diagnostic center. By implementing Fuzzy inference system an attempt has been made in this work to detect heart disease.

An agent based consultancy system has been given in Figure 3 concerning to the patient. Here, both the symptoms and the parameters of the disease is uploaded by the patient into the server using the method of hospital registration. The data uploaded are then passed to the server which communicates with the diagnostic center for testing the symptoms. If the diagnostic parameters are in the normal range then remedial steps has to be taken.

Figure 2. Diagnostic centre structure

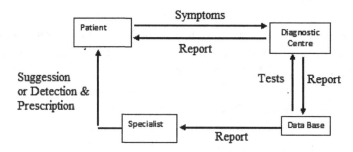

Figure 3. Information centre

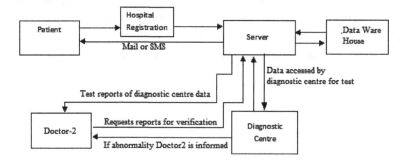

In case of any abnormalities, the diagnostic center informs the specialist or Doctor 2 for further course of action. The fuzzy IF-THEN rules have been used here to provide the desired information to the corresponding doctor. The doctor then verifies the medical reports by sending a request for the same to the server for further investigation of prescription.

The diagnostic agent problem can be formulated as below:

Using the FIS for the adaptive cardiac disease detection, we have used these following parameters.

- The range of BP is 90 to 190(in mm Hg)
- The range of Serum Cholesterol is 120-560 (Mg/dl)
- FBS 60-140 (Mg/dl)
- ECG ST-T wave abnormality

These above mentioned inputs have been chosen due to the following.

- When BP is normal the heart function can be considered normal. With rise in BP, the abnormally in heart occurs which indicates there is a chance of cardiac disease. Hence, it tends to be an important parameter for heart disease detection.
- The blood circulation path narrows due to the deposition of TG. This is due to the presence of excess Cholesterol in blood. Ultimately the circulatory path gradually narrows and can cause heart disease in a patient.
- Sugar happens to be a silent killer. It increases the fatality in a patient with heart attacks in case the level is beyond the normal value.
- ECG signal happens to be the best predictor of heart ailment. It is indicated by studying the ST-T wave. Any abnormality in this wave indicates the potential heart disease even though the other symptoms mentioned earlier are normal.

In this work, our goal is to consider the experience of the physician in order to store those in a set of fuzzy tables. The inference engine is used to develop a computer program which can at its own find out whether a patient is affected with some specified symptoms with certainty from the listed suspected diseases. A crisp percentage value is assigned to this certainty for each suspected disease. The four parameters are considered for the diagnosis using Fuzzy inference system is fuzzified.

Electrocardiogram

For the fuzzification of electrocardiogram three fuzzy sets has been considered such as normal, ST-T abnormal and hypertrophy. The MFs of both normal and hypertrophy fuzzy sets have been considered trapezoidal whereas the MF of ST-T abnormal has been triangular.

Blood Pressure

Out of systolic and diastolic blood pressure parameters we have considered Systolic blood pressure in our study and different input variables has been divided into three fuzzy sets which are low, normal and high. The membership functions for low, normal and high has been taken as trapezoidal, triangular and trapezoidal respectively.

Fasting Blood Sugar

The fasting blood sugar is also one of the important factors for abnormality detection and therefore we have also considered in our investigation. We have defined the range of blood sugar value into three, i.e., low, normal and high which has the range below 70, 70–135, 130 and above respectively.

Cholesterol

Medical experts have the opinion that cholesterol has silent effect on the cardiac malfunctioning and can make it worse very rapidly. The total cholesterol a combination of high density lipoprotein (HDL), low density lipoprotein (LDL) and triglyceride plays a significant role to predict the cardiac diseases. Therefore the total cholesterol has been considered for the investigation. Here three fuzzy sets namely less risk, borderline high and high MF has been assigned to the range of values less than 200, 190–245, 240and above respectively. The MFs are taken here as trapezoidal, triangular and trapezoidal respectively.

Defuzzification

The process of conversion of fuzzy variables obtained as output from the fuzzy logic rules into real values is known as defuzzification. The value gives the information about the system and helps to take any action. In Sugeno model, there are two types of defuzzification methods, i.e., weighted average (WA) and weighted sum (WS). In this work, we have used the weighted average method for defuzzification.

Fuzzy Inference

A patient before his diagnosis undergoes a thorough examination. A set of symptoms let S be obtained. The strength of the measurable symptoms can be expressed in terms of numerical values whereas the non-measurable values are expressed in terms of high, medium or low.

Let:

$s[f] ==$ be the fuzzy value for a feature f from the input symptoms.

$r_{ij} = j^{th}$ relevant feature of the ith disease.

$P_{ij}[r_{ij}, v] =$ certainty of presence of the ith disease when the relevant feature r_{ij} has a fuzzy value v.

$\delta_{ij} =$ diagnosis decision of the ith disease based on the relevant feature r_{ij}.

$k_i =$ total number of relevant features for the ith disease.

$w_{ij} =$ weight of the r_{ij} feature in diagnosing the ith disease.

$\sigma_i =$ overall diagnosis decision for the ith disease.

The r_{ij} feature has its own effect upon the diagnosis decision which can be obtained from the profile table $P_{ij}[r_{ij}, v]$ of the disease. For the feature r_{ij} as $s[r_{ij}]$ out of all the symptoms space the fuzzy value v can be obtained. The effect δ_{ij} can get one of the fuzzy sets *Yes*, *May Be*, and *No*. It can be described as follows:

$$\delta_{ij} = P_{ij}[r_{ij}, s[r_{ij}]] \tag{1}$$

By adding all the effects of all k_i relevant parameters, the overall diagnosis decision for the ith disease would be obtained as follows:

$$\sigma_i = \left(\sum_{j=1}^{j=k_i} w_{ij}\delta_{ij}\right) / \left(\sum_{j=1}^{j=k_i} w_{ij}\right) \tag{2}$$

The weight factor w_{ij} used in the above equation plays a major role in decision making. The physician can adjust this value depending upon the importance of a particular symptom while diagnosing for a particular disease. The proper adjustment of the weights reduces the human error in disease prediction. But if all the symptoms carry equal importance during diagnosis then the weight factor is taken as 1. So the equation (5.5) is further simplified to

$$\sigma_i = \frac{1}{k_i} \sum_{j=1}^{j=k_i} \delta_{ij} \tag{3}$$

The crisp values are obtained by specifying the certainty of presence for every disease in the set D. This fuzzy set will be as shown in Figure 3.

The center of area method has been chosen for defuzzification.

Let: c_i represents the centroid of the overall diagnosis decision fuzzy set, c_y is the centroid for the *Yes* fuzzy set and q_i denotes the certainty of presence of the considered disease d_i in percent. Then the decision crisp value for a particular disease d_i can be computed as shown below.

If the results were **yes** for all the related features of d_i, then the decision is considered to be 100%.

$$q_i = (c_i / c_y) \times 100\% \tag{4}$$

Algorithm

10 inputs: Gender, Age, Chest Pain, Blood Pressure, Cholesterol, Blood Sugar, ECG, Heart Beat, Exercise.

1. Output: The heart condition in terms of the linguistic terms.
2. Each chosen input has a set of fuzzy variables
3. Each fuzzy Variable is having its own MF.
4. The MF is computed for every fuzzy variable

Based upon the Mf the rule strength of the fuzzy variable is calculated.

Heart condition is estimated based on the selected output maximum used as the final result.

Proposed Method Using ANFIS

This part of work proposes an ANFIS model to detect and diagnosis of heart related problems. The model uses the internet so as to provide better healthcare facilities. The system comprises of a number of agents who communicates among themselves via internet as shown in Figure 4.

Figure 4. Proposed intelligent health care system

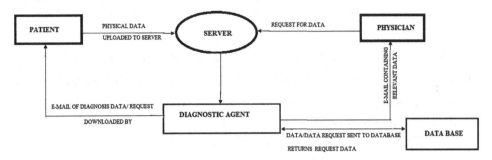

This Figure provides different blocks to represent different agents including the communication paths using solid lines. The patient acts as a user agent with an user id who communicates his/her health parameters or symptoms and the pathological data to a server using a PC or mobile SMS. All these data are uploaded into the server by the patient which is subsequently collected by the diagnostic agent. The uploaded patient data is also stored in the server database for future reference. The analysis of the data is done by the diagnostic agent using the ANFIS technique so as to diagnosis and detect the disease. In case the patient condition is normal, the patient is informed about his status via an email with some precautionary instructions. Similarly, an alert e-mail is sent to the concerned physician in case some abnormality is found in the report for further consultation. The system also provides the facilities of checking the medical history of the using his/her database stored in the server at any time.

Two vital symptoms such as the ECG and BP have been used in this work for diagnosis and detection of the cardiovascular diseases. These symptoms act as the inputs to our proposed ANFIS model. The output or result is the disease identification.

ANFIS Structure

The ANFIS model used in this work has six layers of feed forward ANN and is followed by fuzzy first order Sugeno model. It is a simpler model and is very less dependent on expert knowhow. The model does not need the defuzzification hence is suitable for objective non-linear fuzzy modeling. There are a total of two fuzzy IF-THEN rules, two inputs such as BP, ECG and one output have been used for the heart disease detection. These rules are:

Rule 1: IF BP is X_1 AND abnormality of ECG is Y_1 THEN

$$f_1\left(bp, ea\right) = r_1\left(bp\right) + s_1\left(ea\right) + t_1$$

Rule 2: IF BP is X_2 AND abnormality of ECG is Y_2 THEN
$$f_2\left(bp, ea\right) = r_2\left(bp\right) + s_2\left(ea\right) + t_2$$

Where X_i and Y_i, $i = 1, 2$ represent the fuzzy sets with respect to the inputs BP and ECG respectively. $f_i\left(bp, ea\right), i = 1, 2$ signify the outputs as a linear combination of the corresponding input variables within the fuzzy region that is specified by either rule-1 or rule-2. The parameters r_i and s_i are the design parameters of rule i and are estimated during the process of learning.

A detail function of the ANFIS model and its layers is explained below:

Layer 1 (Input Layer): It is a layer without any sort of computations. The inputs such as BP and ECG data are given to the input nodes. The variables from the input layer moves to the hidden layers.

Layer 2 (Fuzzification Layer): It has a number of adaptive nodes. Each node generates the respective membership grades with respect to the input variable. The outputs of the hidden nodes $\left(O_i^1\right)$ are represented as

$$O_i^2 = w_i = \mu_{X_i}\left(bp\right), \; i = 1, 2 \tag{5}$$

$$O_i^2 = w_i = \mu_{Y_{i-2}}\left(ea\right), i = 3, 4 \tag{6}$$

The membership functions are either continuous or piecewise differentiable functions. The or piecewise differentiable functions are in the form of generalized bell shaped, Gaussian, and triangular. Using a Gaussian function the output of the node (O_i^2) is estimated as

$$\mu_{X_i}\left(bp\right) = e^{-\frac{1}{2}\left(\frac{bp - c_i}{\sigma_i}\right)^2}, i = 1, 2 \tag{7}$$

$$\mu_{Y_{i-2}}\left(ea\right) = e^{-\frac{1}{2}\left(\frac{ea - c_i}{\sigma_i}\right)^2}, i = 3, 4 \tag{8}$$

where c_i and σ_i represent the centers and the width of the Gaussian function respectively. These parameters characterize the fuzzy sets that describe each input variable.

Layer 3 (Rule Antecedent Layer): The layer has the fixed nodes which provides the antecedent part of the associate rule whose firing strength w_i is calculated by the product t-norm operator of each node. The output of each node is represented by

$$O_i^3 = w_i = \mu_{X_i}\left(bp\right) \cdot \mu_{Y_i}\left(ea\right) i = 1, 2 \tag{9}$$

Layer 4 (Strength Normalization Layer): Normalization of the firing strength of layer 3 is accomplished by the fixed nodes of this layer. The output at each node of this layer is given by

$$O_i^4 = \overline{w_i} = \frac{w_i}{w_1 + w_2} i = 1, 2 \tag{10}$$

Layer 5 (Consequent Layer): The layer comprises of adaptive nodes which are computed as a product of the normalized firing strength obtained from the 4th layer. It gives a first order polynomial as

$$O_i^5 = \overline{w_i} f_i = \overline{w_i}\left(r_i\left(bp\right) + s_i\left(ea\right) + t_i\right) i = 1, 2 \tag{11}$$

Layer 6 (Inference Layer): It consists of fixed node which provides the overall output of the given system and is expressed as

$$O_i^6 = \sum_{i=1}^{2} \overline{w_i} f_i = \frac{\sum_{i=1}^{2} w_i f_i}{\sum_{i=1}^{2} w_i} \tag{12}$$

The ANFIS model has two adaptive layers such as layer 2 and layer 5. The eight modifiable parameters are present in layer 2 i.e. {c_i and σ_i: $i=1, 2, 3, 4$}. The input membership functions known as the antecedent parameters are used to communicate among these parameters. Similarly, six modifiable parameters {r_i, s_i and t_i: $i=1,2$} represent the layer 5. The link among these parameters is made using the first order polynomials or the consequent parameters.

Proposed Method Using DNN for Diabetes Detection

In this paper we have used DNN for the classification of the Pima Indian Diabetes data set. In Figure 5 proposed work for diabetes detection is presented.

Deep Neural Network (DNN)

The network has input, hidden and output layers as basic building blocks. It has the potential application in the field of image processing, data mining, pattern recognition, biomedical informatics, speech enhancement, etc. It has different layers for data representation which learn the corresponding representation by expanding the deliberation level between layers. The basic structure has been shown in Figure 6. In the proposed model the training and testing data are chosen from the original data set and is fed to the DNN model to validate the results.

Database

Here in this work the data is taken from UCI repository (Bache & Lichman, 2013). The data set contains total 768 female patient records among them 268 patients are suffering from Diabetes. Here in this record total eight features are taken for the experiment. In Table 1 the total description of the data set has been presented.

Figure 5. Proposed block diagram for diabetes detection

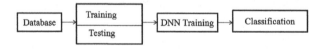

Figure 6. Structure of the DNN

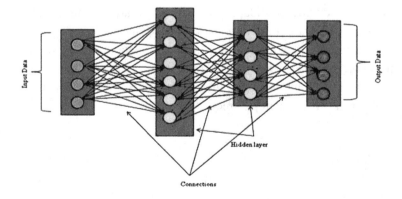

Table 1. Complete description of the dataset

Feature Variable	Minimum Value	Maximum Value	Mean	Median
Pregnant (number of times)	0	16	3.92	2.99
Concentration ofPlasma glucose	0	199	120.89	117.00
Diastolic BP	0	123	68.10	72.00
Thickness of Triceps skin fold	0	98	19.53	23.00
2 hour serum insulin	0	546	79.79	30.5
Body mass index (BMI)	0	66.9	31.00	32.00
Diabetes pedigree function	.078	2.42	0.47	0.37
Age	21	81	33.24	29.00

The data set has been divided into training and testing set with 70% and 30% ratio. Total 534 and 234 data samples are taken for training and testing purpose respectably. Each data set contains class variable 0 for negative and 1 for positive.

Traing of the DNN

The neural network is trained with eight numbers of hidden layers. In this purposed model ReLu (Rectified Linear Unit) activation function is used for the training of the network. ReLU activations are the simplest non-linear activation function and it can be used for the training of the large networks. It allows each unit in the network to express more data in the training of the network. As we know that our data set contains total 8 numbers of features so here the network is trained with eight input layer. In the output of the network it has two layer negative and positive with Softmax activation function. This function is used in the final layer of the DNN based classifier.The softmax function compresses the outputs of each unit to be between 0 and 1. The output for each unit can be compressed within 0 and 1 by softmax function. This function highlights the major value and destroys the smaller one. The yield of this function is proportional to an absolute probability distribution; it discloses to you the probability of the classes is valid.

Classification

In this experiment the classification is done using Testing dataset. Total 234 data samples are taken for the validation of the network to check the patients suffering from Diabetes. The classification result is shown in the result section by the confusion matrix obtained from the Testing data set.

Experimental Setup

The FIS has been developed using the parameters such as the FBS, BP, ECG ST-T abnormality, and Cholesterol and heart disease detection as the output. In Table 2, Linguistic variable versus Fuzzy set for BP and FBS is presented.

The MFs of input using BP and their range:

- **Low:** [40/90 -70/100]
- **Normal:** [70/110-80/120]
- **High:** [90/130 and above]

The MFs of input using FBS and their range:

- **Low:** [60-70]
- **Normal:** [70-120]
- **High:** [120 and above]

The FIS uses the input parameters as ECG, and output as Detection. The set of variable are listed in Table 3.

The MFs of input using ECG and their range:

- **L:** 0
- **H:** 1

The FIS uses input parameters as FBS, ECG, and BP and the output as the Detection and the set of these variables are listed in Table 4.

Table 2. Linguistic variable vs. fuzzy set for FBS and BP

Linguistic Variable	Fuzzy Set, L: Low, N: Normal, H: High
BP	{ L, N, H }
FBS	{ L, N, H }
Detection	{ Yes orNo}

Table 3. Linguistic variables vs. fuzzy set for ECG

Linguistic Variable	Fuzzy Set, L: Low, N: Normal, H: High
ECG	{ L, N, H }
Detection	{ Yes or No}

Table 4. Linguistic variables vs fuzzy set for FBS, BP and ECG

Linguistic Variable	Fuzzy Set, L: Low, N: Normal, H: High
ECG	{Normal, Abnormal}
BP	{ L, N, H }
FBS	{ L, N, H }
Detection	{ Yes or No}

The MFs of input using ECG and their range:

- **Low:** 0
- **Normal:** 0.5
- **Abnormal:** 1

The MFs of input using BP and their range:

- **Low:** [40/90 -70/100]
- **Normal:** [70/110-80/120]
- **High:** [90/130 and Above]

The MFs of input using FBS and their range:

- **Low:** [60-70]
- **Normal:** [70-120]
- **High:** [120 and above]

The FIS has the input parameters as FBS, ECG, Cholesterol and BP and the output is the detection. The set of variable are listed in Table 5.

Table 5. Linguistic variables used with fuzzy set for ECG, BP and FBS

Linguistic Variable	Fuzzy Set, L: Low, N: Normal, H: High
ECG	{ L, N, H }
BP	{ L, N, H }
FBS	{ L, N, H }
Cholesterol	{ L, N, H }
Detection	{ Yes or No}

The MFs of input using ECG and their range:

- **Low:** 0
- **Normal:** 0.5
- **Abnormal:** 1

The MFs of input using BP and their range:

- **Low:** [40/90 -70/100]
- **Normal:** [70/110-80/120]
- **High:** [90/130 and above]

The MFs input using FBS and their range:

- **Low:** [60-70]
- **Normal:** [70-120]
- **High:** [120 and above]

The MFs of input using Cholesterol and their range:

- **Low:** [130]
- **Normal:** [130-230]
- **High:** [230 and above]

RESULT AND DISCUSSION

The input variable Cholesterol, we use Low Density Lipoprotein (LDL) are considered. However, it is also possible to use High density Lipoprotein (HDL). For the input Blood Pressure, the Systolic Blood Pressure is used. The MF remains as an important parameter that represents each fuzzy variable. It determines the rules strength.

- **Age:** This input is described by four fuzzy sets or linguistic variable such as Young, Mid, Old, and Very old. Each variable is associated with a MF. The fuzzy set ranges described the input variable i.e. age has been given in Table 6.
- **Chest Pain:** This input is described in this work using four fuzzy sets such as specific Angina (1), Nonspecific Angina (2), Non-Angina (3), and Asymptomatic (4). At any time, any patient can suffer from only one type of

Table 6. Range of ages

Input Field	Range	Linguistic Representation, Y: Young, M: Mid, O: Old, VO: Very Old
Age	<33 31-47 39- 60 55>	Y M O VO

linguistic variable representing the chest pain. Following numerical has been assigned to denote the particular chest Pain.

- **Cholesterol:** It influences the result more in comparison to other symptoms or input fields. It is either of Low Density Lipoprotein (LDL) or of High density Lipoprotein (HDL). The proposed represented by four fuzzy sets. The range and linguistic representation of input field Cholesterol system only considers the LDL although it is also possible to choose the HDL. This input field has been is shown in Table 7.

- **Gender:** It has been represented by two fuzzy sets such as the Male and the Female. Two numerical has been attached to these fuzzy sets i.e.

  ○ 1 = Male
  ○ 0 = Female.

- **Blood Pressure (BP):** The BP can be Systolic, Diastolic and Mean types. It is another factor that risks the usual heart of human being. The proposed system considers the Systolic BP although any type of BP can be chosen. It is represented by four fuzzy sets such as low, medium, high and very high. The MF is estimated based on the ranges of the Linguistic variable representing the BP and has been shown in Table 8.

- **Heart Rate:** This input field has been represented using three fuzzy sets or linguistic variables such as low, medium and high with the corresponding MFs. Table 9 provides the ranges of each linguistic variable representing the heart rate.

- **Blood Sugar:** This input field often plays a major role that changes the results. Each corresponding fuzzy variable is given a MF according to the

Table 7. Range and linguistic representation of input field cholesterol

Input Field	Range	Linguistic Representation, L: Low, M: Medium, H: High, VH: Very High
Cholesterol	< 196 187-249 218-308 280>	L M H VH

Table 8. The range and the linguistic representation of the input field blood pressure

Input Field	Range	Linguistic Representation, L: Low, M: Medium, H: High, VH: Very High
Blood Pressure	< 132	L
	126- 152	M
	141-171	H
	152>	VH

Table 9. The range and linguistic representation for heart rate

Input	Range	Fuzzy Sets, L: Low, M: Medium, H: High
Heart rate	< 140	L
	112-193	M
	150>	H

range. The ranges and the linguistic representation of fuzzy sets representing the blood sugar have been shown in Table 10.

- **Electrocardiography (ECG):** ECG can be measured using several signal waves drawn in a graph paper. These waves are T wave, Q wave, P wave, S wave corresponding to the electric pulses of heart muscle. In general, these electric waves reside in upper bound of the extracted graph. In case of any wave reside below, and then it indicates some sort of abnormality exists. In the proposed system, the S and the T waves are assumed to be below the upper limit as an indication of abnormality. We have denoted this as ST-T abnormality in the proposed system. The ECG field has been represented here using three fuzzy sets such as Normal, ST-T abnormal and Hypertrophy. The ranges of ECG fuzzy sets have been provided in Table 11.

Table 10. The range and linguistic representation blood sugar

Input	Range	Linguistic Representation
FBS	>=120	Yes(1)
	<120	No(0)

Table 11. The range and linguistic representation of input field ECG

Input	Range	Fuzzy Sets H: Hypertrophy
ECG	<0.5	N
	0.3 - 1.7	ST-T Abnormal
	1.5>	H

- **Output:** The output indicates the on-set of heart disease. It is denoted by an integer value ranging from 0 to 4 that indicates "no presence as 0 to Healthy condition as 4". The increase in this integer value is a sign of level of risk in heart condition. The Output of heart disease is represented by five fuzzy sets such as Healthy, Mild, Moderate, Severe, and Very Severe. It is shown in Table 12 which also provides the output MF ranges.

It has the intelligent agents that will detect the chosen disease. In case of any abnormalities occur, both the patient and the physicians are intimated accordingly. Figure 7 describes the fuzzy rule viewer with these inputs and the predicted Cardio-Vascular disease output.The formation of the rule using the FIS model has been shown in Figure 8.

Table 12. The Range and Fuzzy sets of the Output Field

Output	Range	Fuzzy Sets
Result	< =1 0-2 1- 3 2-4 3>	Healthy Mild Moderate Severe Very Severe

Figure 7. FIS rule viewer for BP, cholesterol, ECG and FBS for detection of cardiovascular disease

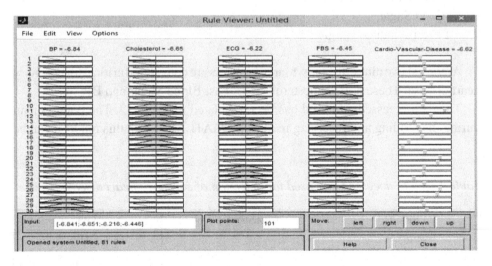

Figure 8. Generated FIS Rule for Cardio-Vascular Disease Detection with Inputs as BP and ECG

```
1. If (BP is LOW) and (ECG is Poor) then (Cardio-Vascular-Disease is Detected) (1)
2. If (BP is LOW) and (ECG is Moderate) then (Cardio-Vascular-Disease is Detected) (1)
3. If (BP is LOW) and (ECG is Good) then (Cardio-Vascular-Disease is Undetermined) (1)
4. If (BP is Medium) and (ECG is Poor) then (Cardio-Vascular-Disease is Undetermined) (1)
5. If (BP is Medium) and (ECG is Moderate) then (Cardio-Vascular-Disease is Not-detected) (1)
6. If (BP is Medium) and (ECG is Good) then (Cardio-Vascular-Disease is Not-detected) (1)
7. If (BP is High) and (ECG is Poor) then (Cardio-Vascular-Disease is Detected) (1)
8. If (BP is High) and (ECG is Moderate) then (Cardio-Vascular-Disease is Detected) (1)
9. If (BP is High) and (ECG is Good) then (Cardio-Vascular-Disease is Detected) (1)
```

Detection of Coronary Heart Disease

The chosen data set for this work has been procured from University of California (UCI), Irvine C.A. centre for machine learning and intelligent systems. The data from 303 patients have been chosen for this piece of experiment. It comprises of seventy-six attributes. Out of thirteen attributes available for experiment of the coronary heart disease, nine has been selected for our purpose. These are as follows:

1. Sex
2. Age
3. Cholesterol
4. Blood Pressure
5. Electrocardiogram (ECG)
6. Type of Chest Pain
7. Maximum of Heart Rate
8. Blood Sugar
9. Exercise induced angina

Among above nine attributes, four parameters are considered critical for coronary heart disease. These are: Cholesterol, ECG, Fast Blood Sugar and BP.

The data accessed and used has been tabulated in Table 13. The popup menu during the sending and receiving mail using MATLAB. From this report it can be

Table 13. Data accessed and used for detection of coronary heart disease

BP	Cholesterol	FBS
94/163	403	251
91/142	352	192

diagnosed. As the data shown it appears to be the patient is suffering from cardiac problem.

The proposed model and experimental set up are used for testing purpose. The measure challenge in this frame work is information system; testing result is to be coordinated so that the exact result can be communicated. Agent technology is employed in this architecture and can be used in e-healthcare system. The method has been accurate in rendering the information to the patient. It has the ability to adapt different information in the current scenario. The model involves a number of agents who work together to resolve and coordinate among themselves for the specific task. Also this system can maximize the performance with consistency at a short notice. The result of detection can be communicated to the patient and the physician for further course of action.

The modifiable parameters of ANFIS model discussed in this work have been used for optimization to facilitate the better cardiovascular disease detection and are shown in Table 14.

The training data for modeling the ANFIS is shown in Figure 9 using the ECG and BP features of 104 patients. There are three membership functions in each input. These are poor, good and moderate for ECG and low, medium and high for BP. Using a number of rules, these membership functions provide cardio vascular disease as the outputs.

Table 14. The detection of cardiovascular disease using BP and ECG symptoms with ANFIS

Sl. No of Patient	Test		Fitness Range		Abnormality		Result	Suggession
	BP	ECG	BP	ECG	BP	ECG		
1	65-110	Bad	80-130	Good	Abn	Abn	Identified	Physician Consultancy
2	62-106	Moderate	80-130	Good	Abn	Abn	Identified	Physician Consultancy
3	70-115	Good	80-130	Good	Abn	Nor	Ambiguous	Risk free
4	80-130	Bad	80-130	Good	Nor	Abn	Ambiguous	Risk free
5	85-120	Moderate	80-130	Good	Nor	Nor	notIdentified	Risk free
6	80-140	Good	80-130	Good	Nor	Nor	notIdentified	Risk free
7	90-140	Bad	80-130	Good	Abn	Abn	Identified	Risky
8	95-145	Moderate	80-130	Good	Abn	Abn	Identified	Risky
9	110-160	Good	80-130	Good	Abn	Nor	Identified	Physician Consultancy

*'Abn' stands for abnormality and 'Nor' stands for Normal

Figure 9. Plot of training data for BP and ECG using ANFIS

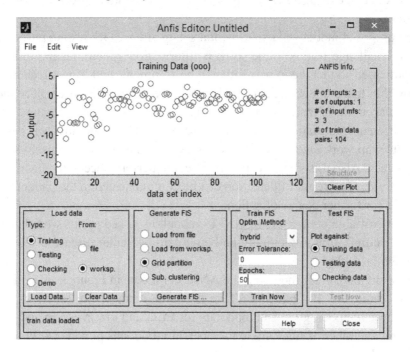

The corresponding FIS output has been shown in Figure 10. In simulating the ANFIS model, Grid partitioning has been used to divide the data with 50 numbers of epochs for training. Figure 11 shows the formed rules for BP and ECG for detection of cardio vascular disease.

The Diabetes data has been obtained from the UCI repository and then we have checked for the missing value. The data set contains 768 patient's records without any missing value. In Figure 12 the data set structure is presented. The data set is divided into Training and Testing set with the probability of 70% and 305 respectably. The Network is trained with the 534 numbers of patients sample. The training performance is presented in Figure 13.

In this experiment the training accuracy is around 100%. The total number of epochs is 300, batch size is 32 and the validation split is 0.2. The loss in this experiment is 0.037. After successfully training of the network the next step is the validation step. The validation has been done with the previously stored 234 testing data set. In Table 15 confusion matrix that we have get is presented. Here we are getting the average classification accuracy 98.5%. For calculation of the classification accuracy we can take the ratio of the accurately classified cases to the total number of cases.

After the detection of the disease, the information is forwarded to different entities via the internet or the mobile network or any of the smart communicating devices.

Figure 10. Plot of FIS output for BP and ECG using ANFIS

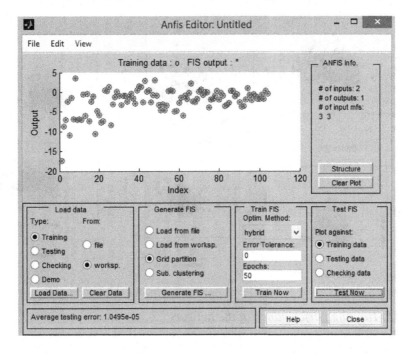

Figure 11. ANFIS rule formation for BP and ECG for detection of cardio-vascular disease

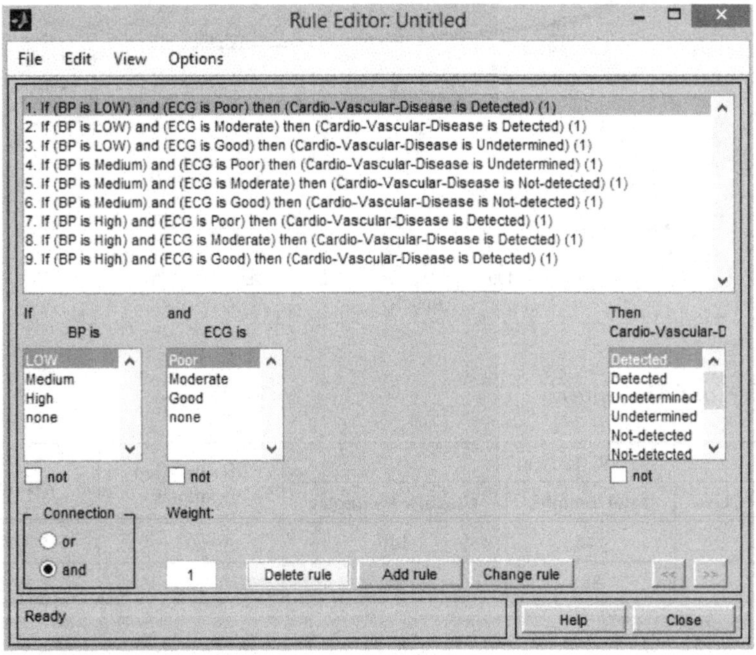

Figure 12. Diabetes dataset structure

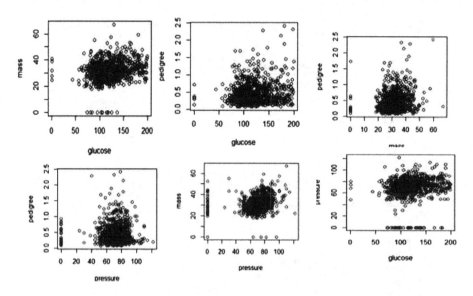

Figure 13. Training performance of the network

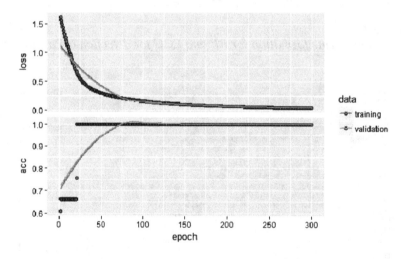

Table 15. Confusion matrix

	Reference		Misclassified Samples	Accuracy
Diabetes Class	**Total Samples**	**Classified Samples**		
Positive	146	146	0	100%
Negative	88	86	2	97%

Average Accuracy= 98.5%

These communicating modules are efficient and highly reliable hence are widely popular in these days. Short Message Service or SMS is one such communicating media that is instant accessible to almost everyone in today's world. Now a day's the mobile and smart devices are associated with the internet technology and with different APPs that facilitates both the patients and the doctors. Sometimes the computer terminal needs a modem for these services (Oresko et al., 2010; Forkan, Khalil & Tari, 2013; Adeli & Neshat, 2010). The proposed work has been verified with both set up with the resultsobtained.The corresponding SMS service applying GSM module is provided in Figure 14.Based ondifferent inputs of a patient, the patient is considered normal and his/her status is indicated as OK.

Effect of the Internet on Patients

With growing internet facilities there have been a corresponding increase in amount of patients. Further, healthy personal are also relying on the internet to monitor their health. This makes the physicians, medical practitioners as well as health care providers to answer question the patients receive from internet blogs. A few of these internet inputs are excellent and advanced. Thus, the physicians can learn from the information obtained by patients in case they are unable to access the internet due to busy schedule. They must be open and transparent to different types of questions

Figure 14. Snapshot of SMS received from the server

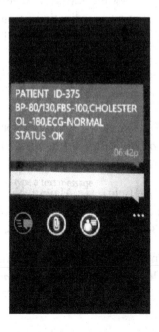

that help for better clinical practices. Further, there are few sites which can allow access to personalized advice with some fee which the physicians cannot ignore for better health care.

CONCLUSION

Theories and techniques of the cognitive sciences provide inputs on many issues that help in the design and implementation of health facilities, information and mobilization. These concepts play a pivot role to enhance a clinician's performance with better patient care with a range of cognitive tasks. In this chapter we have presented two distinct work on decision support system based on DNN architecture for heart disease diagnosis and Fuzzy based model for cardiac problem analysis. The experiment performs well with the proposed algorithm and is validated with standard UCI repository dataset. Diabetes is one of the long term diseases and it needs proper treatment at very early stage. If a machine can correctly classify the disease with the information then it can help doctors for better treatment. Deep learning has established to be a proficient technique for the analysis of the large volume of data. Here in this paper we have applied DNN based classification for the women affected by Diabetes. Pima Indian Diabetes data set has been taken for experiment. This work emphasizes on the monitoring and detection of cardio vascular disease using the Fuzzy based model for smart e-health care. This model will help both the patient and the physician by providing the desired information on patient health condition. Based on this information coordination among the hospital management, doctor and the patient can be materialized for future diagnosis and follow-up.

REFERENCES

Adeli, A., & Neshat, M. (2010, March). A fuzzy expert system for heart disease diagnosis. In *Proceedings of International Multi Conference of Engineers and Computer Scientists (Vol. 1)*. Academic Press.

Al Rahhal, M. M., Bazi, Y., AlHichri, H., Alajlan, N., Melgani, F., & Yager, R. R. (2016). Deep learning approach for active classification of electrocardiogram signals. *Information Sciences, 345*, 340–354. doi:10.1016/j.ins.2016.01.082

Al-Sakran, H. O. (2015). Framework architecture for improving healthcare information systems using agent technology. *International Journal of Managing Information Technology, 7*(1), 17–31. doi:10.5121/ijmit.2015.7102

AlSharqi, K., Abdelbari, A., Abou-Elnour, A., & Tarique, M. (2014). Zigbee Based Wearable Remote Healthcare Monitoring System for Elderly Patients. *International Journal of Wireless & Mobile Networks, 6*(3), 53–67. doi:10.5121/ijwmn.2014.6304

Bache, K., & Lichman, M. (2013). *UCI machine learning repository*. University of California, School of Information and Computer Science. Retrieved from http://archive.ics.uci.edu/ml

Biswal, S., Das, A., Shalinee, S., & Mohanty, M. N. (2016). Design of telemedicine system for brain signal analysis. *International Journal of Telemedicine and Clinical Practices, 1*(3), 250–264. doi:10.1504/IJTMCP.2016.077918

Biswal, S. S., Mohanty, M. N., Das, S., & Sahu, B. (2016). Unconscious state analysis using intelligent signal processing techniques. *Advanced Science Letters, 22*(2), 314–318. doi:10.1166/asl.2016.6856

Chi, C. L., Street, W. N., & Katz, D. A. (2010). A decision support system for cost-effective diagnosis. *Artificial Intelligence in Medicine, 50*(3), 149–161. doi:10.1016/j.artmed.2010.08.001 PMID:20933375

Craven, M. W., & Shavlik, J. W. (1997). Using neural networks for data mining. *Future Generation Computer Systems, 13*(2-3), 211–229. doi:10.1016/S0167-739X(97)00022-8

Das, R., Turkoglu, I., & Sengur, A. (2009). Effective diagnosis of heart disease through neural networks ensembles. *Expert Systems with Applications, 36*(4), 7675–7680. doi:10.1016/j.eswa.2008.09.013

Devi, C. S., Ramani, G. G., & Pandian, J. A. (2014). Intelligent E-healthcare management system in medicinal science. *International Journal of Pharm Tech Research, 6*(6), 1838–1845.

Elavathingal, E. E., & Sethuramalingam, T. K. (n.d.). A Survey of EHealthcare Systems Using Wearable Devices and Medical Sensor Networks. *KJER, 1*(1).

Forkan, A., Khalil, I., & Tari, Z. (2013, September). Context-aware cardiac monitoring for early detection of heart diseases. In *Computing in Cardiology Conference (CinC), 2013* (pp. 277-280). IEEE.

Forkan, A., Khalil, I., & Tari, Z. (2013, September). Context-aware cardiac monitoring for early detection of heart diseases. In *Computing in Cardiology Conference (CinC), 2013* (pp. 277-280). IEEE.

Hannan, S. A., Mane, A. V., Manza, R. R., & Ramteke, R. J. (2010, December). Prediction of heart disease medical prescription using radial basis function. In *Computational Intelligence and Computing Research (ICCIC), 2010 IEEE International Conference on* (pp. 1-6). IEEE.

Haykin, S. (1999). Neural Network (2nd ed.). Academic Press.

Jan, B., Farman, H., Khan, M., Imran, M., Islam, I. U., Ahmad, A., ... Jeon, G. (2017). Deep learning in big data Analytics: A comparative study. *Computers & Electrical Engineering*. doi:10.1016/j.compeleceng.2017.12.009

Liu, W., Wang, Z., Liu, X., Zeng, N., Liu, Y., & Alsaadi, F. E. (2017). A survey of deep neural network architectures and their applications. *Neurocomputing, 234*, 11–26. doi:10.1016/j.neucom.2016.12.038

Lokanath, S., Narayan, M. M., & Srikanta, P. (2016). Critical Heart Condition Analysis through Diagnostic Agent of e-Healthcare System using Spectral Domain Transform. *Indian Journal of Science and Technology, 9*(38). doi:10.17485/ijst/2016/v9i38/101937

Mohanty, M. N. (2016). An Embedded Approach for Design of Cardio-Monitoring System. *Advanced Science Letters, 22*(2), 349–353. doi:10.1166/asl.2016.6854

Mohapatra, S. K., Palo, H. K., & Mohanty, M. N. (2017). Detection of Arrhythmia using Neural Network. *Annals of Computer Science and Information Systems, 14*, 97–100. doi:10.15439/2017KM42

Nahar, J., Imam, T., Tickle, K. S., & Chen, Y. P. P. (2013). Association rule mining to detect factors which contribute to heart disease in males and females. *Expert Systems with Applications, 40*(4), 1086–1093. doi:10.1016/j.eswa.2012.08.028

Ordonez, C. (2006). Association rule discovery with the train and test approach for heart disease prediction. *IEEE Transactions on Information Technology in Biomedicine, 10*(2), 334–343. doi:10.1109/TITB.2006.864475 PMID:16617622

Oresko, J. J., Jin, Z., Cheng, J., Huang, S., Sun, Y., Duschl, H., & Cheng, A. C. (2010). A wearable smartphone-based platform for real-time cardiovascular disease detection via electrocardiogram processing. *IEEE Transactions on Information Technology in Biomedicine, 14*(3), 734–740. doi:10.1109/TITB.2010.2047865 PMID:20388600

Osowski, S., & Linh, T. H. (2001). ECG beat recognition using fuzzy hybrid neural network. *IEEE Transactions on Biomedical Engineering, 48*(11), 1265–1271. doi:10.1109/10.959322 PMID:11686625

Ozcift, A., & Gulten, A. (2011). Classifier ensemble construction with rotation forest to improve medical diagnosis performance of machine learning algorithms. *Computer Methods and Programs in Biomedicine, 104*(3), 443–451. doi:10.1016/j.cmpb.2011.03.018 PMID:21531475

Pal, D., Mandana, K. M., Pal, S., Sarkar, D., & Chakraborty, C. (2012). Fuzzy expert system approach for coronary artery disease screening using clinical parameters. *Knowledge-Based Systems, 36*, 162–174. doi:10.1016/j.knosys.2012.06.013

Rahman, Q. A., Tereshchenko, L. G., Kongkatong, M., Abraham, T., Abraham, M. R., & Shatkay, H. (2015). Utilizing ECG-based heartbeat classification for hypertrophic cardiomyopathy identification. *IEEE Transactions on Nanobioscience, 14*(5), 505–512. doi:10.1109/TNB.2015.2426213 PMID:25915962

Ram, R., & Mohanty, M. N. (2017). The Use of Deep Learning in Speech Enhancement. *Annals of Computer Science and Information Systems, 14*, 107–111. doi:10.15439/2017KM40

Sarangi, L., Mohanty, M. N., & Patnaik, S. (2016, June). Design of ANFIS Based E-Health Care System for Cardio Vascular Disease Detection. In *International Conference on Intelligent and Interactive Systems and Applications* (pp. 445-453). Springer.

Sarangi, L., Mohanty, M. N., & Patnaik, S. (2017). Detection of abnormal cardiac condition using fuzzy inference system. *International Journal of Automation and Control, 11*(4), 372–383. doi:10.1504/IJAAC.2017.087045

Sarangi, L., Mohanty, M. N., & Pattanayak, S. (2016). Design of MLP Based Model for Analysis of Patient Suffering from Influenza. *Procedia Computer Science, 92*, 396–403. doi:10.1016/j.procs.2016.07.396

Sarangi, L. N., Mohanty, M. N., & Patnaik, S. (2016). Cardiac Diagnosis and Monitoring System Design using Fuzzy Inference System. *Rice (New York, N.Y.)*.

Sikchi, S. S., Sikchi, S., & Ali, M. S. (2013). Design of fuzzy expert system for diagnosis of cardiac diseases. *IJMSPH*. Retrieved from: http://citeseerx.ist.psu.edu/viewdoc/download?doi=10.1.1.301.1840&rep=rep1&type=pdf

Silverman, B. G., Hanrahan, N., Bharathy, G., Gordon, K., & Johnson, D. (2015). A systems approach to healthcare: Agent-based modeling, community mental health, and population well-being. *Artificial Intelligence in Medicine, 63*(2), 61–71. doi:10.1016/j.artmed.2014.08.006 PMID:25801593

Singh, M., Mohanty, M. N., & Choudhury, R. D. (n.d.). An Embedded Design for Patient Monitoring and Telemedicine. *International Journal of Electrical, Electronics, and Computer Engineering, 2*(2), 66-71.

Su, C. J. (2008). Mobile multi-agent based, distributed information platform (MADIP) for wide-area e-health monitoring. *Computers in Industry, 59*(1), 55–68. doi:10.1016/j.compind.2007.06.001

Tupe, S., & Kulkarni, N. P. (2015). *ECA: Evolutionary Computing Algorithm for E-Healthcare Information System.* Retrieved from SPPU, Pune iPGCON-2015: http://spvryan. org/splissue/ipgcon/063. pdf

Yan, H., Zheng, J., Jiang, Y., Peng, C., & Xiao, S. (2008). Selecting critical clinical features for heart diseases diagnosis with a real-coded genetic algorithm. *Applied Soft Computing, 8*(2), 1105–1111. doi:10.1016/j.asoc.2007.05.017

Zhang, Q., Yang, L. T., Chen, Z., & Li, P. (2018). A survey on deep learning for big data. *Information Fusion, 42*, 146–157. doi:10.1016/j.inffus.2017.10.006

Chapter 2
A Cognitive Information Retrieval Using POP Inference Engine Approaches

Parul Kalra
Amity University, India

Deepti Mehrotra
Amity University, India

Abdul Wahid
Maulana Azad National Urdu University, India

ABSTRACT

The focus of this chapter is to design a cognitive information retrieval (CIR) framework using inference engine (IE). IE permits one to analyze the central concepts of information retrieval: information, information needs, and relevance. The aim is to propose an inference engine in which adequate user preferences are considered. As the cognitive inference engine (CIE) approach is involved, the complex inquiries are required to return more important outcomes as opposed to customary database questions which get irrelevant and unsolicited responses or results. The chapter highlights the framework of a cognitive rule-based engine in which preference queries are dealt with while keeping in mind the intention of the user, their performance, and optimization.

DOI: 10.4018/978-1-5225-5793-7.ch002

INTRODUCTION

An ample amount of data is being generated on a daily basis by various modern communications and computer systems. Data repositories with Internet are considered to be the most effective and commonly used, for prevailing information in time as well as space (Václav Snášel, 2010; Gorrab et al., 2016). The data on the internet is accessible by everyone, irrespective of their age. The information retrieved from the net is not as exigent as it can be because of deficient resources and low performance. Information retrieval (IR) also called information storage and retrieval or information organization and recovery is the craftsmanship and study of recovering from an accumulation of things, a subset that fills the client's need. Cognition is the procedure by which the tangible information is changed, decreased, explained, put away, recouped, and used. Cognitive IR or Cognitive Information Retrieval (CIR) is a technique for real world communication between the user and search engine that has to be maintained so that the required information is available to the user in the desired form keeping in mind the optimization and performance criteria. When some data is saved by a person in the storehouse, it is expected that any other person can retrieve that same data.

Expert systems (Inference Engine) are regarded with preference for enacting huge information for any field of expertise. Further, with the reasoning the problems can also be solved.

BACKGROUND

An information retrieval (IR) system returns the information to the user based on the query that has been given. Typically, an IR system searches in bulk of data which can either be unstructured or semi-structured. The essentiality of the IR system urges due to rapid increase in the corpus, where the traditional systems fail to apply. The amount of information stored grew rapidly from 1956 to 2005 and is still increasing. Today, in this digitalized world and with the high speed networks, enormous amount of data is being collected. The only possible way of retrieving relevant information is via IR systems.

CIR OPTIMIZATION

CIR optimization is based on collective information of below:

- Previous user activities (Preferences)

- Search criteria (Optimization)
- Information models (Performance)

Refer to Figure 1 CIR with IE and Performance, Optimization and Performance. The purpose of the optimization is to aggrandize the retrieved records as per the request of the end-user even when the user has ambiguity on the information that is to be looked upon (Václav Snášel, 2010; Gorrab et al., 2016). The important characteristic for an Inference Engine (IE) that has to be taken care off is its efficiency and performance. Therefore, it is necessary to research the methods that make an efficient IE that can give exact results in minimum possible time. The structure should ensure the working with huge volumes of data. Eventually these data are kept in CIR databases. The information is available in many forms such as text, photos, videos, etc. Joining different parts of this information with the queries, gives a valid reason for the displayed result. But, this result can lead to the contradictions and inconsistencies for different users with respective preferences, even though they might be searching for the same thing. The working of the engine depends upon the architecture and logic used in making the application. For the improved retrieved data portrayal and for understanding the query in a better way using preferences, the advanced CIR techniques (Semantics and ontology, Information retrieval, Semantic Web, Context, Cognitive engineering, Statistical correlation measures) (Carsten Keßler, 2010) are being studied. Even though, optimization and performance characteristics are still missing in them. Customization of the searches done on the web, modifying user

Figure 1. CIR with IE and performance, optimization and performance

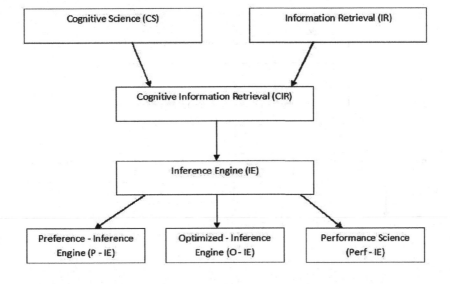

profile according to the use and soft information retrieval is the ultimatum (Balke et al., 2002). The objective of this paper is to bring out those search queries that relate to the above mentioned tasks. A framework is being explained, which will help the user to search the records as per the prior searches made by him. Also, its implementation has been discussed that will help the user to derive the utmost pertinent data (current context) from the internet (Václav Snášel, 2010) keeping in mind the optimization of queries and performance of the system. So, we are introducing an inference engine i.e. Cognitive POP Engine. Refer to Figure 2 for the Block diagram of Cognitive Information Retrieval.

Case Study: Where We Are and What We Want?

The real parts of the human data gathering procedures are oblivious, users are not always aware of what they actually want. Still try to express them to whatever extent they can like in a non-verbal communication. If the interaction is between two humans, one can still try to understand the other through his actions, expressions etc. But in-case of human- computer interaction, it is not very effective (Ackoff et al., 1972). consider a case when a person is searching for something such as live premier league scores then the preference engine would search according to personal profiles, intention, circumstance and space using a rule base of cognitive heuristics

Figure 2. Block diagram of cognitive information retrieval

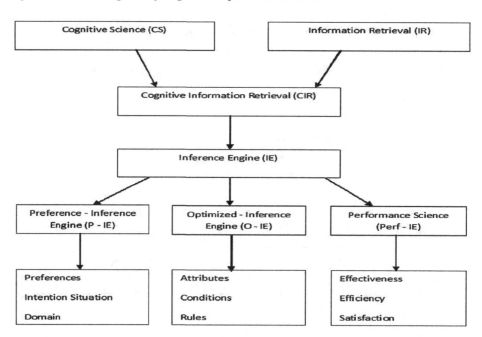

but the results we get may or may not be optimized as the search topic may not be put as it is required for exact searching (Balke et al., 2002) if the engine is optimized then it can guess the correct structure to be searched from the entered one and can sequence the query correctly. But engine loses time while structuring the query hence, the performance of the system decreases. The efficiency and effectiveness of the query we get from the preferences and optimizations sometimes leads to lack of satisfaction as it may require more time for expansion. Hence it is important to keep up the performance of the system so that the engine can have all the three attributes- *preference, optimization* and *performance* at the same time. The engine will learn with its persistent use and the search results will be optimized irrespective of the request structure and results will be obtained quickly and that too according to the situation and profile of the user. In the current scenario of live premier league scores, the Indian user with a cricket history will get the Indian premier league as the first link, the Bangladesh user with the same history will get Bangladesh premier league as the first link, an English fan with cricket history will get county cricket and football history will get Barclays premier league link as the first. Even the muddled requests will be handled by the engine with optimization feature. The past usage, situation and domain of the user will be taken in order to give better search results and the query can be entered in any pattern, but the results will be accurate and quick. This is how the proposed cognitive pop engine is expected to work. Refer Figure 2 block diagram of cognitive information retrieval.

COGNITIVE POP (PREFERENCE, OPTIMIZATION AND PERFORMANCE) ENGINE: A PEEP VIEW

The design of features like preference, optimization and performance (POP) are usually related to an individual directly; be it the manner he behaves, his perception or the way he explores the information. These features hold the exploratory process. It comprises the retrieval of relevant records and supports the intellectual activities (such as sense-making) that are related with the searches made (Kelly, 2009; Fakhfakh et al., 2016). The specific user interest and his understanding come from his preference and expectation. The system that is being used now-a-days is bounded to the set of keywords given by the user and thus, they hold a disadvantage of not having an insight of the user (Balke et al., 2002; Bhatia et al., 2013). We have categorized the cognitive POP Engine into three phases:

P-Inference Engine

The P-Inference Engine chooses what inclinations are to be decided for the extension relying upon individual profiles, intention, current state and domain with the help of a rule based on cognitive heuristics. This data can be accumulated from (Balke et al., 2002):

- Long-term preferences
- Intention
- Situation
- Domain

O-Inference Engine

A rule-based inference engine has the following essential elements (Griffin et al., 1989):

- **Attributes:** X1, X2, ..., Xn1
- **Conditions:** C1, C2, ..., Cn2
- **Rules:** R1, R2, ..., Rn3
- **Actions:** A1, A2, ..., An4

Performance Inference Engine (PIE)

PIE has the ability to carry out the conducts like popup alarms, writing messages in system log, and initiating programs as an answer to the stated query other than computing arithmetic and logical values. These conducts are very advantageous for detecting, tracking and rectifying the information (Ruthven, 2005). The basic components of performance are:

- Effectiveness
- Efficiency
- Satisfaction

COGNITIVE POP ENGINE: ARCHITECTURAL DESCRIPTION OF P-INFERENCE ENGINE

The task of fetching stored data as per the query made by the user even in his "abnormal condition of learning" is information retrieval (Kelly, 2009). The type of information needed by an individual changed from one person to another according to different scenarios like the background of the person. This need can be characterized as:

- **Verificative:** First the properties of the documents are identified then according to them the data is sorted. Properties such as name of the author, title etc.
- **Conscious Topical:** The topic is identifiable & defined, yet, less correct than in the main classification. A man searching for data has some level of conception of it.
- **Muddled or Ill-Defined Needs:** It includes those scenarios where an individual fails to convey what he actually wants to search for; he might be searching for some new concept (Ingwersen et al. (1995)).

Experts are subject authorities in their errand space, rather than pros in documentation, for example, online inquiry specialists. Further, the experts of any particular field are distinguished from the non- professionals. For instance, in a bank the individuals employed for keeping the records, accounts etc. are considered to be the professionals of that field (Morten Hertzum et al., 2012; Fakhfakh et al., 2016). Most existing IRS puts forward a quick retrieval but this method is quite expensive as far as there updates are concerned (Faloutsos, 1985). In spite of these expensive updates it is still famous because the updates are limited and centralized in contrast to the retrieval. According to our objectivity, IRS for the experienced requires greater variability (Codd, 1970). We will draw a clearer picture of this by considering the architecture and the by putting IRS into effect based on preference. Thus, efficient retrieval holds more importance than capability to handle changes. Displaying client inclinations has turned into a basic piece of recovery in databases and data frameworks (Hristidis et al., 2001; Stober, 2017) and formal inclination designing has increased much consideration. To design the inclinations, the priority is to be set first. Either to the topic of how client inclinations can be reached upon or to replying to the inference (Balke et al., 2002).

As a rule, today's frameworks offer two potential outcomes:

- Clients need to express their inclinations unambiguously keeping in mind the end goal to alter an administration or to represent a question.
- The framework gets inclinations naturally from prior connections of a similar client or he gets some parts of the client to a run of the mill use profile. Refer to Figure 3 for the architecture of Cognitive Information Retrieval

Figure 3. Architecture of cognitive information retrieval using POP ENGINE

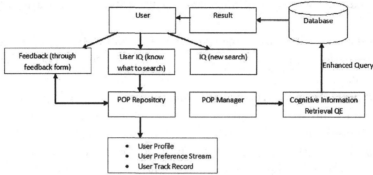

To associate with clients, two interfaces can be valued: notwithstanding a customary inquiry interface a visual displaying instrument for complex inclinations is required that comprehends the associations between inclinations expressed. An interdisciplinary cognitive research is required for deriving preferences in order to establish a suitable rule base as looked forward by the usage patterns, domains, situations etc.

- **Long-Drawn Precedence:** The idea of significance from past retrievals is utilized.
- **Objective:** Motivation for retrieving, of a particular client is incorporated into ventures.
- **Situation:** The current state and milieu of a client is utilized.
- **Domain:** Information on a particular space is utilized.

"Cognitive database retrieval along with utilizing the straightforward search terms provided by the user will also make use of long drawn specifications of usual likes and dislikes (if applicable), the specific intention for the search (purpose directed), the personal situation and environment (location, time, emotions, etc.)". Along with specific learning, the accumulation of client concerning data appears to be encouraging to characterize a particular thought of importance and empowers broad displaying of recovery. In this manner the client is bolstered in inquiry undertakings and even the development of client learning in every space can be considered by versatile client profiles containing increasingly any particular data (Balkeet al. (2002)) thus improving its performance in evaluating same or similar query when asked again. The upcoming in this research will focus on how modelling, picking along with joining inclinations can be improved for better optimization and performance (Balke et al., 2002; Gorrab, 2016). Refer to Figure 4 for the General Architecture of Cognitive IR.

Figure 4. General architecture cognitive information retrieval

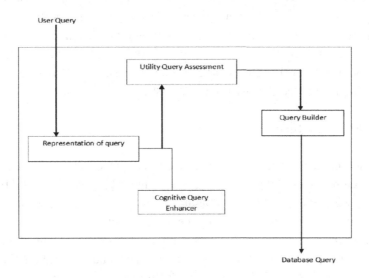

Optimal Inference Engine

The information that is needed behind Query formulation is ill-defined making which is non-trivial. In an easier language, countless ways can be used to express a single idea or subject. In order to retrieve only relevant data from the storehouse as per the queries, we should be able to identify all the keywords explaining that particular information which is to be retrieved. Thus, we are obscuring the many-to-many relationship between subjects and expressions, or concepts and words by sticking to words.

The rules of optimization can be seen in a particular order. It can be expressed as:

```
if <condition statement> then <statement 1>
```

The above line says that "if" a condition statement is true "then" the corresponding statement 1 will be executed.

After the inspection done by the inference rules,the actions are executed if the information supplied by the user fulfils the condition statements. *Forward chaining* and *backward chaining* are 2 methods of inference.

1. **Forward Chaining:** This technique takes the facts as and when accessible and then tries to come up with a final decision. The actions that are being executed is the result of the prior step. Also, it is a top to down technique. An example of this technique is Expert.

2. **Backward Chaining:** As the name suggests, it is a bottom to top technique. It initiates with the goals and the queries that satisfies the conditions mentioned in the rules. This technique is not an exploration technique, rather it is a verification technique. Example of this method is MYCIN. A system which uses both backward and forward chaining is Prospector. Performance Engine chaining is Prospector (Griffin et al., 1989).

Performance Engine

Most of the classical IR measures were assessed and established under various retrieval circumstances along with some beliefs regarding the views and behaviors about relevance. This was a major problem. The assumptions underlie the developed evaluation measures thus questioning the applicability and usefulness of these measures. There is a considerable variation in the actual relevance measuring technique. Number of studies have been examined and studied (**1**) about the concept of relevance (Borlund, 2003; Mizzaro, 1997; Ruthven, 2005; Saracevic, 2007) (**2**), criteria employed by the users while making relevance assessment (Tombros et al., 2005) and relevance measuring technique (Toms et al., 2004; Vakkari et al., 2004; Kek'al'ainen et al., 2002; Eisenberg, 1988) .

The classic way of performance assessment has some difficulty. The difficulty is in the ability to gather binary relevance assessment. Also, it is necessary that the documents that are the most relevant should appear in the top results as the user does not consider the data which is down in the list to be of much use. We have to make sure that the users do not have to go through any effortful task so as to find the relevant documents. Moreover, if huge amount of irrelevant data appears on the top than the subject is likely to get deviated from this topic that he was searching for.

Time plays a very critical role as far as the IR evaluations are considered. The roles are played at both, gross and specific level. Consider the time taken by the user for searching any particular task, it is the gross level instance and the time taken by the user because of getting engaged searching for any particular task is the specific level example. Although, the time taken by the users for searching a particular topic is a measure that examines the efficiency of the IR but the time taken by the machine is also a way of measuring the performance of IR. It significantly contributes in evaluating the systems performance.

- *Effectiveness* is the "accuracy and completeness with which users achieve specified goals." Or we can say that, a system is effective only if it is capable to accomplish the tasks that the user wants to achieve.

- *Efficiency* is the "resources expended in relation to the accuracy and completeness with which users achieve goals." Or a system is effective if the users are able to complete their desired task, added the expense and the efforts should be as minimal as possible.
- *Satisfaction* is the "freedom from discomfort, and positive attitudes of the user to the product" (Ruthven, 2007). Precisely, it is the satisfaction attained by fulfilling a specific task. The satisfaction is felt by the user when he is able to retrieve the data he was actually searching for (Diane Kelly, 2009)).

CONCLUSION

The notion for the advanced personalization, optimization and performance of the retrieval system has been studied in detail in our paper. The inclination should be towards all the sub-parts of the personalized information systems. The forthcoming architectures are expected to ensure the presence of adequate interfaces for modeling and preference input (Balke et al., 2002), Spink et al.(2005)) along with optimization and performance. However, psychological experiments show that even with complex preference queries (Balke et al. (2002), Stober, S. (2017)), difficulties will be faced by the user thus making optimization and performance difficult to manage. Hence, the use of cognitive technique by the systems should be encouraged to support users in stating their profiles (Balke et al., 2002; Gorrab et al., 2017) and to optimize the entered queries so as to fulfill the exact search requirement of the user at considerable speed. The objective of the cognition query expansion is to put into effect the queries that are knowledge dependent, which permits a complex query to give quick and accurate results. The results obtained from the queries along with the involvement of cognitive knowledge are much more relevant as compared to the results from the queries that were used earlier (Balke et al., 2002). In addition to the improved proficiency in customization (preference modelling, optimization and performance), the consequences from sociology and psychology are also to be emphasized in the retrieval procedure. Cognitive database retrieval along with the use of explicit user-provided search terms, will also use long-drawn specification of general precedencies and dislikes, the time, emotions, etc.). Specific learning this accumulated client-related data appears to be encouraging to characterize a particular thought of importance and empowers a broad displaying of recovery inclinations a long way past the consider question terms gave and at a higher rate. In this manner the client is bolstered in inquiry undertakings and even the development of client learning in every space can be considered by versatile client profiles containing increasingly particular data (Balke et al., 2002) thus, improving its performance in evaluating same or similar query when asked again. The upcoming in this research will emphasize on the way of modeling, picking and joining inclinations (Balke et al., 2002) with better methods of optimization and performance. As the applications in utilization designs for the web administrations have earned encouraging outcomes (Wagner et al., 2002) utilizing inclination and enhancement alone, assist change should be possible to develop the zone of utilization and show signs of improvement comprehension of psychological procedures amid recovery assignments with these elements (Balke et al., 2002). The features such as the path which is absolutely necessary to be traversed is explored, quick response, relevant response can be added to the existing systems using the cognitive POP engine.

REFERENCES

Ackoff, R. L., & Vergara, E. (1981). Creativity in problem solving and planning: A review. *European Journal of Operational Research*, 7(1), 1–13. doi:10.1016/0377-2217(81)90044-8

Balke, W. T. (2002). A roadmap to personalized information systems by cognitive expansion of queries. In *Proc. of the EnCKompass Workshop on Content-Knowledge Management and Mass Personalization of E-Services (EnCKompass 2002)* (pp. 123-125). Academic Press.

Bhatia, P. K., Choudhary, C., Mehtrotra, D., & Wahid, A. (2013). A review of the cognitive information retrieval concept, process and techniques. *Journal of Global Research in Computer Science*, 4(3).

Borlund, P. (2003). The concept of relevance in IR. *Journal of the Association for Information Science and Technology*, 54(10), 913–925.

Codd, E. F. (1970). A relational model of data for large shared data banks. *Communications of the ACM*, 13(6), 377–387. doi:10.1145/362384.362685

Eisenberg, M. B. (1988). Measuring relevance judgments. *Information Processing & Management*, 24(4), 373–389. doi:10.1016/0306-4573(88)90042-8

Fakhfakh, R., Feki, G., Ammar, A. B., & Amar, C. B. (2016, October). Personalizing information retrieval: A new model for user preferences elicitation. In SMC (pp. 2091-2096). Academic Press.

Faloutsos, C. (1985). Access methods for text. *ACM Computing Surveys*, 17(1), 49–74. doi:10.1145/4078.4080

Gorrab, A., Kboubi, F., & Ghezala, H. (2017). Social Information Retrieval and Recommendation: state-of-the-art and future research. *HAL*. Retrieved from: https://hal.archives-ouvertes.fr/hal-01444570/document

Gorrab, A., Kboubi, F., Ghezala, H. B., & Le Grand, B. (2016, November). Towards a dynamic and polarity-aware social user profile modeling. In *Computer Systems and Applications (AICCSA), 2016 IEEE/ACS 13th International Conference of* (pp. 1-7). IEEE.

Griffin, N. L., & Lewis, F. D. (1989, October). A rule-based inference engine which is optimal and VLSI implementable. In *Tools for Artificial Intelligence, 1989. Architectures, Languages and Algorithms, IEEE International Workshop on* (pp. 246-251). IEEE.

Hertzum, M., Søes, H., & Frøkjær, E. (1993). Information retrieval systems for professionals: A case study of computer supported legal research. *European Journal of Information Systems*, *2*(4), 296–303. doi:10.1057/ejis.1993.40

Hristidis, V., Koudas, N., & Papakonstantinou, Y. (2001). Prefer. *Proceedings of the 2001 ACM SIGMOD international conference on Management of data - SIGMOD 01*. 10.1145/375663.375690

Ingwersen, P., & Willett, P. (1995). An introduction to algorithmic and cognitive approaches for information retrieval. *Libri*, *45*(3-4), 160–177. doi:10.1515/libr.1995.45.3-4.160

Kekäläinen, J., & Järvelin, K. (2002). Using graded relevance assessments in IR evaluation. *Journal of the Association for Information Science and Technology*, *53*(13), 1120–1129.

Kelly, D. (2009). Methods for evaluating interactive information retrieval systems with users. *Foundations and Trends® in Information Retrieval, 3*(1–2), 1-224.

Keßler, C. (2010). *Context-aware semantics-based information retrieval*. IOS Press.

Mai, J. E. (2016). *Looking for information: A survey of research on information seeking, needs, and behavior*. Emerald Group Publishing.

Mizzaro, S. (1997). Relevance: The whole history. *Journal of the Association for Information Science and Technology*, *48*(9), 810–832.

Ruthven, I. (2005). Integrating approaches to relevance. *New Directions in Cognitive Information Retrieval, 19*, 61-80.

Ruthven, I., Baillie, M., & Elsweiler, D. (2007). The relative effects of knowledge, interest and confidence in assessing relevance. *The Journal of Documentation*, *63*(4), 482–504. doi:10.1108/00220410710758986

Saracevic, T. (1975). Relevance: A review of and a framework for the thinking on the notion in information science. *Journal of the Association for Information Science and Technology*, *26*(6), 321–343.

Saracevic, T. (2006). Relevance: A review of the literature and a framework for thinking on the notion in information science. Part II. In Advances in librarianship (pp. 3-71). Emerald Group Publishing Limited.

Saracevic, T. (2007). Relevance: A review of the literature and a framework for thinking on the notion in information science. Part III: Behavior and effects of relevance. *Journal of the Association for Information Science and Technology*, *58*(13), 2126–2144.

Snášel, V., Abraham, A., Owais, S., Platoš, J., & Krömer, P. (2010). User profiles modeling in information retrieval systems. In Emergent Web Intelligence: Advanced Information Retrieval (pp. 169-198). Springer London.

Spink, A., & Cole, C. (2005). A multitasking framework for cognitive information retrieval. *New directions in cognitive information retrieval*, 99-112.

Stober, S. (2017). Toward Studying Music Cognition with Information Retrieval Techniques: Lessons Learned from the OpenMIIR Initiative. *Frontiers in Psychology*, *8*, 1255. doi:10.3389/fpsyg.2017.01255 PMID:28824478

Tombros, A., Ruthven, I., & Jose, J. M. (2005). How users assess web pages for information seeking. *Journal of the Association for Information Science and Technology*, *56*(4), 327–344.

Toms, E. G., Freund, L., & Li, C. (2004). WiIRE: The Web interactive information retrieval experimentation system prototype. *Information Processing & Management*, *40*(4), 655–675. doi:10.1016/j.ipm.2003.08.006

Vakkari, P., & Sormunen, E. (2004). The influence of relevance levels on the effectiveness of interactive information retrieval. *Journal of the Association for Information Science and Technology*, *55*(11), 963–969.

Chapter 3
Assessing Animal Emotion and Behavior Using Mobile Sensors and Affective Computing

Heath Yates
Kansas State University, USA

Brent Chamberlain
Utah State University, USA

William Baldwin
Biosecurity Research Institute, USA

William Hsu
Kansas State University, USA

Dana Vanlandingham
Biosecurity Research Institute, USA

ABSTRACT

Affective computing is a very active and young field. It is driven by several promising areas that could benefit from affective intelligence such as virtual reality, smart surveillance, perceptual interfaces, and health. This chapter suggests new design for the detection of animal affect and emotion under an affective computing framework via mobile sensors and machine learning. The authors review existing literature and suggest new use cases by conceptual reevaluation of existing work done in affective computing and animal sensors.

DOI: 10.4018/978-1-5225-5793-7.ch003

INTRODUCTION

Affective computing is a young field that large industry is starting to take notice of. For example, Facebook has recently unveiled a suicide prevention application that relies on big data and machine learning to detect the potential for suicide (Constine, 2017). Facebook goes beyond traditional sentiment analysis in examining content posted by users by trying to infer a potential emotional state that might be dangerous to its users. In addition, Microsoft has developed powerful software that can detect emotions based on facial expressions (Parkinson, 2015). The potential for human centric applications is showing a lot of promise. In the context of affective computing and animal sensors, it is not much of a leap to consider the potential of an animal centric affective computing. Animal agriculture in the United States, specifically livestock and poultry, is a major industry in the United States with past and projected annual exports often exceeding $25 billion dollars every year (Cooke, Jiang, & Heerman, 2017). At the very least, an animal agricultural approach to affective computing might be an area of fast-untapped potential. The authors believe an explicit approach to study core affect shared by humans and animals carries advantages that will be discussed later. In short, the authors believe there is rich merit to pursue affective computing in animal sensors.

Figure 1. Intensive animal farming
© 2017, SmartStock. All Rights Reserved. Used with Permission.

One of the challenges in writing on an interdisciplinary topic is addressing properly the broad collection of themes. In our case, the specific focus will be on affective computing, animal sensors, human, and animal affect. Therefore, the authors wish to caveat our approach by saying that this chapter is not attempting to be exhaustive in our coverage of ideas, but painting a broad and clear picture. The authors believe that affective computing and animal sensors are still very young and nascent fields. As such, the authors believe the time has come for the explicit and deliberate pursuit of animal affect via affective computing based on animal sensors. In fact, much of the groundwork already exists and the authors desire to show that an affective computing approach to animal sensors is viable and requires an explicit conceptual reevaluation of what animal sensors can do by making affect a core research goal. This chapter shall first proceed by discussing the historical development and contemporary state of affective computing. Second, it shall briefly give similar coverage to animal sensors. Third, the chapter will put affective computing and animal sensors into a broader context by discussing human and animal affect. The chapter will tie these together by discussing an affective computing approach to animal sensors and a potential design for affect and arousal detection using animal sensors.

AFFECTIVE COMPUTING

In this section, the authors will provide a brief discussion of affective computing as motivation for its implications in animal sensor research. This chapter shall proceed in a chronological order outlining the intellectual origins of affective computing. As this chapter progresses, it shall outline some of the ever present challenges and future promises of the field. Finally, this chapter will briefly touch upon some potential

Figure 2. Empatica E4 Sensor

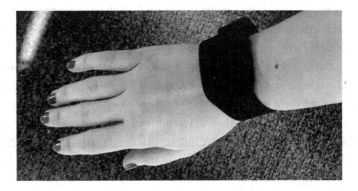

that is still unexplored in animal sensor research, which will be explored in detail in the last section of our chapter.

Affective computing is the study and implementation of intelligent systems, computational devices, and wearables that can recognize, interpret, and process human affect (Picard 1997). Affective computing is still a relatively young and nascent field (Soleymani et al., 2017). However, it can trace it's origins back to the early 1960s when Page proposed that users would someday be able to interface and share their thoughts and feelings real time with machines (Page, 1962). Popular imagination was not far behind with Arthur C. Clarke's Hal 9000 computer which was popularized in Kubrick's 1968 film "2001: A Space Odyssey". The imagined machine had impressive capabilities such as recognizing individuals, reading lips, playing chess, art critique, and recognizing emotions expressed by his human companions (Picard, 2001). It was only in the mid-1990s when affective computing was formally proposed (Picard, 1995). However, the dream of having intelligent software agents was quickly proposed and investigated (Picard and Rosalind, 2000). It was also proposed that machine intelligence must include emotional intelligence (Martinez-Miranda & Aldea, 2005).

Early work in affective computing was focused on outlining fundamental approaches and capabilities. Since its inception, there has been a particular focus on wearables and statistical learning algorithms. Specifically, the examination of custom affective wearables that look for patterns in biometrics such as heart rate, blood pressure volume, and galvanic skin response as well as, core affect such as smiling, angry gestures, strained voices (Picard and Healey, 1997). It did not take long for the researchers to demonstrate the viability of processing digital affective signals with statistical learning for pattern recognition such as Fishers linear discriminant and Hidden Markov Models (Healey and Picard, 1998; Riseberg, Klein, Fernandez, & Picard, 1998). Skin conductivity signal was shown to be a statistically significant indicator of being startled and correlated with arousal level (Healey and Picard, 1998). Devices such as glasses to detect interest or confusion in a human face via the muscles were demonstrated as well (Scheirer, Fernandez, & Picard, 1999). Research into core human affect such as neutral, anger, hate, grief, platonic love, romantic love, joy, and reverence was first explored in an affective computing context using digitized electromyography, blood volume pressure, galvanic skin response, and respiration waveforms (Vyzas and Picard, 1999). The early years of affective computing were notable for their rapid exploration of using wearables, raw biometrics signals, advanced statistical techniques to detect and recognize affective responses. Obviously, some of the core human affect emotions is beyond the scope of animal sensors and animal science, but the biometric and wearable centric approach for animals is worth emulation.

The next decade demonstrated even more rapid advances. During this time the first applications to healthcare emerged. The implications of having affective medicine study frustration, irritation, stress, and health were explored (Picard, 2002). Research demonstrated the viability of detecting stress and heart health with a phone and wearable computer (Picard and Du, 2002). In addition, humans statistically rated higher software that was designed to simulate caring over software that did not (Bickmore and Picard, 2004). Health applications were explored by providing computers that interact with health patients with relational skill such as empathy, social dialogue, and nonverbal immediacy behaviors led to statistically significant higher proactive viewing of health information (Bickmore, Gruber, & Picard, 2005). Intelligent systems that systematically analyzed voice patterns and assisted individuals diagnosed with autism and down syndrome found patterns in their speech (Hoque, 2008). This early work was perhaps instrumental in the development of more evolved affect systems for autistic individuals that allowed participants play video games which support their social development (Khandaker, 2009). Stress detection also continued to make advances in the context of drivers by using physiological signals collected in a car during driving (Healey and Picard, 2005). Facial recognition software began to emerge for the detection of human affect by focusing on tracking eyes and eyebrows in real time and using PCA analysis (Kapoor and Picard, 2002). Speech analysis for stress and affect was applied to driver's speech using dynamic Bayesian networks and mixture hidden Markov models (Fernandez and Picard, 2003). The examination of skin conductance in an affective computing context made further advances by exploring mobile sensor applications via Bluetooth technology, demonstrating the ability of mobile wearables to detect emotional and physical arousal of users (Strauss et al., 2005). Several of the advances explored above would establish the baseline and viability of commercial affective applications that would later come, some by the same researchers, which shall be discussed in a more later. Ethical considerations briefly came to the forefront after a study conducted by DARPA investigated emotional state recognition to determine potential criminal intent (Reynolds, 2004). In addition, adversarial applications of affective computing were explored in a poker game setting (Reynolds, 2005). Minimalist interface design was first promoted for affective systems as well (Wren and Reynolds, 2004). In addition, typography has been shown to influence affect and has potential in affective systems (Larsons and Picard, 2005). A notable study of affective computing was the proposal of a multi-sensor affect recognition system based on multimodal Gaussian Process approach to detect interest or disinterest in children (Kapoor and Picard, 2005). This is important because of the explicit emphasis on the multimodal approach to affect detection. A slight extension of this Gaussian process approach was used to explore prediction of student frustration and have a virtual tutor intervene (Kapoor, Burleson, & Picard, 2007). An important study examined the viability of a low-cost, light, mobile and unobtrusive electrodermal activity (EDA) for wristband detection of affect (Poh, Swenson, and Picard, 2009; Poh, Swenson, & Picard, 2010). The work between 2000 and 2010 was notable for the expansion of affective computing into several contexts such as health, education, driving, and gaming. Advances in mobile computing and sensors in general led to increasingly sophisticated experimental affect systems which became multimodal in nature. Lastly, many applications and advances in sensors would set the stage for the commercialization. However, the authors believe the viability of commercialization for the detection of affect in animals will be much more immediate and in parallel to exploration of affective computing systems in an animal sensor context.

Recent progress has a synergy about it that makes us feel that society is on the cusp of the next big breakthrough (Waldren, Agresta, & Wilkes, 2017). Advances in mobile sensor technology and increasing computing capabilities of mobile computing have not gone unnoticed in affective computing. The potential and ability of mobile health wearables and mobile phones to collect long-term continuous physiological signals such as electrodermal activity, plethysmography for the potential treatment and management of physical, neurological, and mental issues has been proposed (Fletcher, Poh, & Eydgahi, 2010). A very novel affective computing system was proposed by using a medical mirror for non-invasive health monitoring of health patients (Poh, McDuff, & Picard, 2011). This user centric approach was further evolved and developed to allow users with a wearable system that gathers both physiological responses and visual context via a cellphone camera to consume both their external affect and internal physiology (Hernandez, Mcduff, Fletcher, & Picard, 2013). More advances to the detection of stress occurred, a machine learning centric approach to stress recognition via cell phone usage while also using wearable sensors that record continuous physiological signals (Sano and Picard, 2013; Sano and Picard, 2014). This is also notable because the methods for detection started to evolve in sophistication with an interest in machine learning. The remote measurement of cognitive stress via heart rate variability was also explored using a multimodal sensor approach (McDuff, Gontarek, & Picard, 2014). The machine learning approach was further evolved in detecting and managing stress via the use of using supervised learning methods (Hernandez, 2015). Around this time, commercialization and products began to appear. For example, the Empatica E3, the predecessor to the more sophisticated and advanced Empatica E4 appeared. The Empatica E3 was a wearable, wireless multi-sensor device that allowed for real-time acquisition of biofeedback data that could be used in the detection and prediction of affect (Garbarino, Lai, Bender, Picard, and Tognetti, 2014). The last decade saw rapid expansion of research into commercial products. In addition, machine learning centric approach to affective learning also has been gaining serious traction. It is likely that human health, fitness, smart homes, and education will continue to show great promise. Also, the trends in big data, data science, and machine learning will likely keep progress consistently ongoing. Notably, the recent approaches by deep learning continue to show promise. In addition, new areas for affective computing centric study are continually being proposed. For example, an affective computing centric approach to the study of built environments has been proposed (Yates, Chamberlain, Norman, & Hsu, 2017). Therefore, the authors believe the commercial and machine learning centric approach in affective computing is worthy of emulation in the pursuit of studying animal affect via sensors.

The authors note that there are few caveats which should be addressed because we cannot cover all aspects of this field due to the rapid pace and evolution of the science; therefore, this chapter is not exhaustive and there are topics which are beyond the scope of this discussion. As can be readily seen, pace in affective computing has been historically rapid. If is important to note here that the field is still in its early stages. While research has been productive and certainly successful, like all endeavors there are challenges. First, it is important to examine the big picture and identify the true scope and aims of the field. Simply said, the larger, overreaching, goal is to make machines emotionally intelligent. In other words, the goal of having an affectively intelligent machine such as the Hal 9000 computer in fiction is likely to still be very far off in the future. Challenges to affective computing that persist to this day were first outlined and fleshed out within the first decade of the field (Sloman 1999; Picard 2003; Tao and Tan, 2005). They essentially boil down as follows: First, emotional expression is so broad with so many complexities that it is an open question on how to best measure them. Second, there are a lot of emotions and affect that are extremely variable across individuals. Thus, it is therefore difficult to measure. Third, how can researchers best apply the sensor and multimodal centric approach to best detect primary emotions or core affect (Sloman 1999, Picard 2001). In addition, more contemporary challenges of how cross cultural issues impact the detection of affect as well as standardization approaches as commercialization increases (Schuller 2017). The authors believe it is important to be cognitive of these challenges in context of affective computing centric approach to animals and sensors. However, the authors also believe that there are some advantages to affective studies for animals over human centric research that this chapter will discuss later in the last section.

Let us briefly consider again the implications of this section as it relates to animal sensors and animal affect. The authors believe the historical backdrop and rapid development of affective computing is a strong motivating force that similar gains can be made in the study of animals and their affective responses. Especially in organisms more closely related to us such as mammals. The authors propose that affect in animals can be detected through sensors collected real time and continuous physiological data. This chapter will now turn its attention towards animal sensors while keeping the current state and capabilities of affective computing in mind. Ultimately, we hope researchers will consider emulating the approach and successes discussed above but for animal sensors.

ANIMAL SENSORS

Before this chapter discusses the potential applications of animal centric affective computing via sensors, there is merit in first discussing animal sensors from a high overview. First, this section shall briefly discuss the definition, goals, and capabilities of animal sensors. Second, it will discuss the historical development and timeline of advances with animal sensors. Next, it will briefly outline the impressive and growing capabilities of contemporary animal sensors as well as the sensors potential for studying animal affect.

By definition, a sensor is a device that can detect or measure a physical parameter and records, indicates, or responds to it (Dictionary, 2007). The capabilities of sensors used with animals is still in its early stages. Consequently, the authors believe the potential for research in animal sensors and affect is still relatively unrealized. Research and development of sensors has traditionally been focused on robotics, military, industrial production, and entertainment (Helwatkar, Riordan, & Walsh, 2014). However, the authors believe this is quickly changing as emerging capabilities and technologies in Internet of Things (IoT), big data storage, and machine learning are allowing previous unrealized potential in animal health and science. For example, in 2011, Fujitsu introduced GyuHo "Cow Step", a system that used Wi-Fi connected pedometers to track the health and reproductive cycles of dairy cows (Monma, 2011). In addition, as the global population is likely to increase, it provides ever increasing incentive to more carefully manage and plan non-sustainable resources, health monitoring, and disease control of animals.

Animal sensors have their origins in the 1980s. One of the earliest uses of sensors was animal identification. There are primarily two approaches with animal sensors. The first approach is to attach the sensor to the animal. This can either be external on the animal or invasive by being in the animal, commonly something that can be swallowed or surgically implanted. The second approach is an external sensor not attached, which can remotely monitor and record data on the animal. One of the first and common uses for external sensors is the measurement for continuously flow of milk from a cow (Rutten, Velthuis, Steeneveld, & Hogeveen, 2013). Several of the approaches in animal sensors, however, have been attached to the animal and been focused on potential health applications. For example, researchers demonstrated in 1994 that glucose sensors could be implanted in animals subcutaneously to measure

Figure 3. Tag for remote animal identification

the potential for diabetes (Shults et al., 1994). The potential of invasive sensors in ruminant animal's health was demonstrated by measuring PH at 15 minutes over a 44 day period (Mottram, Lowe, McGowan, & Phillips, 2008). In addition, a pilot study has demonstrated that body temperature of a dairy cow can be measured via a rumen bolus (Ipema et al., 2008; Schutz and Bewley 2009). Pedometers have also been demonstrated to be useful in the detection and prediction of oestrus behavior (Løvendahl and Chagunda, 2010; Brehme, Stollberg, Holz, & Schleusener, 2008). Interestingly, the early innovations in sensors in the 1990s were also accompanied by demonstrating the viability of machine learning techniques in determining basic cow behavior (Mitchell, Sherlock, & Smith, 1996).

Contemporary advances in animal sensors have been more rapid. Researchers have used lost-cost GPS sensors and machine learning to detect and accurately predict cow behavior rate (Godsk and Kjærgaard, 2011). Three dimensional accelerometer data and support vector machines have been applied to cow behavioral pattern recognition as well (Martiskainen et al., 2009). Industrial progress in rudiment sensor technology has made significant gains. Current technology is increasingly smaller, longer battery life, has onboard data memory, and is significantly mobile with long range wireless transmission capabilities. SmartStock has a system that records the temperature of ruminant animals using a RF transmitter bolus which can send several data parameters to remote cloud storage.

Bolus technology has led to rapid research in cow behavior and physiology such as ovulation detection, understanding on how water consumption lowers core body

Figure 4. Rudimen Temperature Bolus Sensor

temperature, associations between udder health, and reproductive performance (Hudson, Bradley, Breen, & Green, 2012). The sensor above is example of how growing long-range sensing capabilities of animal sensors provide researchers and industry previously unrealized potential in studying animal behavior and health.

Since the inception of animal sensors, approaches have been primarily on animal behavioral, health, and disease implications. However, the capabilities and potential of animal sensor is nascent. Currently, research is still unimodal but that should and is beginning to change as the cost and sophistication of sensors and technologies continue to decrease. The future of animal sensors research will be increasingly multimodal. The methodologies for examining the data is also growing in sophisticated in parallel as advances in machine learning are happening at a record pace. For example, innovative machine learning approaches are starting to explore the visual space to detect and predict animal behavior (Nasirahmadi, Edwards, & Sturm, 2017). There has been an implicit interest in the affective state of the animal, but as far as the authors are aware it has not been very explicit in the agricultural animal space. In general, the work and interest in affect for animals still appears to be very nascent. The current brief examination of animal sensors is by no means exhaustive and such a coverage would be beyond the scope of this chapter. Nevertheless, the brief coverage here demonstrates that the continuing evolution in sensors capabilities, big data storage, and machine learning methodologies suggest that an affective computing approach centered on sensors and machine learning in studying animal affect and behavior has arrived.

Figure 5. Solar Powered Telemetry Receiver

ANIMAL AND HUMAN AFFECT

This section will now present a discussion concerning animal and human affect. The first portion of this section is devoted to keeping our eye on the ball, so to speak, by being mindful of the goals of affective computing while considering the challenges. After this, we will discuss, for comparative purposes, how it bears some merit to consider how other fields, such as artificial intelligence, have dealt with their challenges by looking to nature for both inspiration and simplification of experiments. For example, modeling mathematically the simple behavior of a neuron has inspired artificial neural networks and deep learning which has been a wildly successful approach to solving a complex problem. Lastly, we will wrap up the discussion by examining the benefits and advantages of shared human and animal affect.

Since the inception of affective computing, research has shown promise in the detection of primary and core human emotions (Picard, Vyzas, & Healey 2001; Poria, Cambria, Bajpai, & Hussain 2017). This chapter has also established in a previous section how a multimodal approach can be used successfully. However, challenges remain which emerged early on in the study of affective computing and have persisted to this day (Sloman, 1999). First, there is the debate regarding the nature of emotions and whether it is most appropriate to measure affect as continuous or as an ordinal measurement (Russell 2017). The discussion of this point will take us into philosophical as well as scientific considerations and is therefore beyond the scope of this chapter. However, it is still important to be cognitive of this challenge and keep it ever in mind during our discussion of animal and human affect. The second challenge in studying human affect is adequately detecting primary and secondary emotions (Kruijff et al., 2017; Sloman, 1999; Picard, 2003). Research has shown that a multimodal sentic approach to the detection of primary emotions is possible and successful (Bosch 2015; Poria, Cambria, Bajpai, & Hussain 2017; Picard, Vyzas, & Healey 2001). As the authors have discussed in a previous section and above, it seems to be that detection of primary affect and emotions has had it successes. The challenge affective computing faces which is to adequately detect complicated and subtle emotions is a complicated problem which persists to this day (Ngo, See, & Phan 2017; Ward & Marsden 2004; Sloman, 1999). Therefore, the scope of this section is limited to the consideration of primary shared affect in animals and humans.

Since accurately measuring an affective response remains the ultimate goal and challenge, it is worth a discussion of how a cousin of affective computing, artificial intelligence (AI), has effectively dealt with its own challenges. It is worth noting that even at the inception of affective computing, unlike AI, researchers noted the difficulties would be a long term endeavor (Picard, 1995; Picard, 1998). Nevertheless, there are several notable parallels. Both fields have lofty goals and both fields have serious challenges. These are hard problems. In addition, affective computing is becoming more commercial and industries may become discouraged once the low hanging fruit has been harvested and the harder problems remain, as was the case with AI. In addition, similar to the study of affective computing, there was the period of initial exuberance and enthusiasm. Thus, this section shall now briefly discuss artificial intelligence in an effort to avoid some of the pitfalls that could befall affective computing.

Artificial intelligence (AI) is the study and design of computer systems which is intelligence displayed by machines in contrast with natural intelligence displayed in humans and other animals (Russell, Norvig, & AI, 1995). Our discussion will also avoid the AI effect and avoid the understanding that AI is "whatever hasn't been done yet" (Hosfader, 1980). Since it's inception, AI has made great advances in natural language processing, vision recognition, and problem solving abilities found in gaming such as Go, Chess, and Checkers. That said, the early years of artificial intelligence was notable for its enthusiasm and exaggerated claims that intelligent machines were nearly round the corner (Minksy, 1967). The initial goal was the creation of machines that would parallel, equal, or surpass human intelligence. The advent of the Mark I Perceptron in 1960 was a major breakthrough that demonstrated that machines could learn by trial and error based on a simplistic neural network that simulates biological learning (Rosenblatt 1958). Popular fiction itself quickly caught on with dreams of super advanced machine intelligences by 2001 such as the fictional HAL 9000 computer (Clarke, 2016). However, the enthusiasm was replaced by disappointment. By the 1980s, the difficulties became notable, research began to slow, and some even called it the winter for artificial intelligence research (Hendler, 2008). Since then pragmaticism has taken hold, focusing on using AI to solve specific problems. For example, the early successes with neural networks evolved into deep learning which have led to great advances in vision recognition (LeCun, Bengio, & Hinton, 2015). Despite the new successes and hypes, advances in hardware and large data sets are unlikely to lead to another winter in AI from occurring in the near future (Knight 2016). Rather than directly try to model human level intelligence in machines, contemporary projects focus on insect, small mammal, and dog level intelligence (Bonabeau, Dorigo, & Theraulaz, 1999; Shannon, 1953; Bohrer, 2012). Also, the pragmaticism remains with researchers acknowledging that AI has still yet to achieve even the level of intelligence that even a rat is endowed with (Shead, 2017). In summary, the authors propose a similar pragmatic and focused approach for the study and design of intelligent affective systems by focusing on shared affect between animals and humans.

As shown in the previous section in this chapter, there have been significant gains and breakthroughs in affective computing based on primary core emotions detection, healthcare applications, and affective facial software recognition. Like artificial intelligence challenges, the authors believe the road to truly sophisticated and affectively intelligent systems is still a long one and that affective computing is still a young field. Therefore, there is merit on focusing to keeping things simple, such as the detection of basic arousal in specific contexts such as built environments (Yates, Brent, Norman, & Hsu 2017). Just like the emphasis of artificial intelligence studies on modeling insects, rats, and other small mammals. The authors also believe that studying animal affect can both be instructive and useful in reducing complexity. This chapter does not wish to overstate the authors' case and readily acknowledge that affective computing research in both animals and humans has always been implicitly present in research (Cummins et. al 2017). However, considering the challenges of the field, the authors believe the focus on shared animal and human affect should be much more explicit and deliberate. Thus, focusing on basic core affect that is shared between mammals and humans is worth consideration.

As this chapter has briefly outlined above and in an earlier section, the goals of affective computing are centered around the development of intelligent systems and wearables that can recognize, interpret, and understand human affects (Picard, 1995; Tao & Tan, 2005). Researchers define primary and core affect as something valanced as either pleasant or unpleasant, dependent upon the pleasure system, and rooted in the evolutionary system (Kringelbach & Berridge 2017). Research indicates the primary unconditioned affect and core affect responses in human and animals as seeking, rage, fear, lust, care, grief, and play (Panksepp, 2011). Based on this, the authors propose initial research in shared animal and human affect by further simplifying this to detectable biophysical arousal states such as comfort, pleasure, and well-being state when in good health (Panksepp 2011). For ethical considerations in research, the authors propose current and future interest in affective states that are not likely to induce fear. It is probably evident that even by simplifying the problem, it is still a complicated endeavor. Ultimately, the persistent challenge of affective computing boils down to measuring affect and providing a metric to subjective experiences.

Neuroscience offers several insights into similarities and differences in human and animal affect (Panksepp, 2004). While the authors acknowledge the immense challenges of measuring emotional states and determining metrics to identify emotional changes, there is clear evidence for similarities of fundamental and primal affective responses such as fear, rage, panic, and seeking (Panksepp, 2011; Panksepp, 2004). The discovery of these emotional similarities, offer an opportunity to compare how different environmental triggers might influence both animals and humans. In highly controlled contexts, where animals and humans are kept under close watch or confinement there can be specific environmental triggers that exacerbate emotional responses. For instance, Kuma and Ng (2001) provide evidence that in psychiatric wards crowding may be associated with risk and feeling of lost safety. The result of overcrowding then leads to patients experiencing less privacy and control within their environmental context (Kuma and Ng, 2001). In this same way, cattle who are confined will also feel more stressed (Mader, 2003), likely with an increase in stress because they lose control of their environmental surroundings. To counteract this effect in humans, the design of therapeutic spaces can mitigate stressful experiences (Connellan et al., 2013) and improve behavior.

AFFECTIVE COMPUTING APPROACH TO ANIMAL SENSORS

First, as discussed in previous sections, there are challenges that have persisted with the evolution and growth of the field. One of these challenges is how to best model emotional expression given extensive complexities and unanswered questions about how to best measure them. Second, that affect can be extremely variable across individuals. Third, how to best apply a multimodal centric approach to detect primary affect. Finally, how cross-cultural factors impact the detection of affect and standardization of approaches for commercialization (Schuller 2017). As the authors have stated before, one of the persistent and primary challenges in affective computing is how to generate a metric for affect, given subjective experiences and conditions. The authors believe researchers can greatly reduce many complexities by focusing on core affect that is shared both by humans and animals. In addition, teaching machines to recognize core animal affect given general well-being such as comfort, pleasure, and in good health. The authors do not wish to overstate our claim, but rather build upon previous work done by scientists and industry who have studied such questions for a long time (Fraser and Duncan, 1998; Walker, Trites, Haulena, & Weary, 2011; Goertz et al., 2017). Our approach differentiates this work because this chapter deliberately focus on using affective computing as a framework in seeking answers to such questions. This method requires that researchers teach machines to recognize affect and seek a metric to describe these phenomena. While there is always variation across individuals, animal science provides us with opportunities to identify similarities. Large scale machine learning and big data experiments by scientists and industry demonstrate the viability of detecting and predicting desired animal behavior. For example, Fujitsu and Microsoft can detect both animal estrus and some animal diseases (Klein, 2017). In addition, companies use pedometers to indicate with high probability the best milking times (Koyama et al., 2017). The authors believe that the technology and methodologies already exist in animal science and industry via machine learning and big data to teach machines animal affect on a large scale. Big data approaches with many animals is a realistic expectation in industry and the authors believe individual variation is less serious of a hurdle in animal centric affective computing. The authors believe the focus on shared core affect between animals and humans also completely side steps the cross-cultural issues that are present in more complicated human centric studies. After all, the focus on core animal affect is unlikely to encounter challenges due to living in different cultures.

This section shall now briefly state our comparison to Artificial intelligence. Artificial intelligence as this chapter has discussed earlier has had several springs, summers, and winters. The authors believe this is because the field began with some lofty goals of a fully self-aware machine, but quickly became mired in the complexities. This is not a bad thing, but the early enthusiasm being dampened by the reality has caused several hype and bust cycles (Crevier, 1993). Artificial intelligence has recovered by focusing on simpler biological processes, problems to optimize, and with better realism. Fortunately, affective computing has largely avoided this by being honest and upfront with the challenges. Nevertheless, the field also has similar if not even grander larger long-term goals. First, there is the notion that intelligent machines must also be infused with emotional intelligence before they can be considered intelligent at a broad level (Picard et al., 2004). Second, teaching machines to have the ability to recognize, read, interpret, and predict emotion continues to be daunting. It is beyond the scope of this chapter to discuss the controversy of this goals or long-term viability of them. Suffice it to say that the authors are sympathetic to their aims and believe there is merit in at least trying consistently for the long term. The authors propose that animal centric approach to affective computing can have similar successes as artificial intelligence by both focusing on fundamental shared core affect with animals and trying to model affective intelligence in simpler animals first.

The focus on this chapter has been oriented toward teaching machines to recognize core affect in animals. However, the authors believe there is merit in at least briefly considering the advantages of simulating animal affect in machines as well. Indeed, one can claim that industry has tried to do this already with industrial toys and products (Sharkey and Sharkey, 2011), but the authors contend the purpose was to create the illusion rather than substantive feeling. Instead, the authors believe that approaches such as Brain-Emulating Cognition and Control Architecture (BECCA) which is a long term artificial intelligence project that seeks to create dog like intelligence is where the focus should be (Rohrer, 2012). This project currently does not have an affective computing component. However, the authors believe that ultimately in the future it may be necessary to include affective intelligence to realize its goal in becoming like a dog. In a way, if the leading researchers of affective computing are correct, it might become inevitable that some of the lofty goals within artificial intelligence are realized through an affective computing approach. The authors believe that one of these important steps might be addressed through and animal-centered affective computing approach. These are lofty ideas, but the authors desire to conclude this brief discussion by being much humbler and practical by suggesting that there is merit in focusing on the very hard and complex problems of animal affect, but ultimately a very simplifying one compared to humans.

Research in animal sensors can provide a less expensive and more highly controlled experimental design framework for measuring animal affect, of course heeding to ethical guidelines and related concerns. With animals often having shorter lifespans and far less exposure to different environmental characteristics and experiences than humans, it may be easier to assess how differences in environmental characteristics influence affective responses. An extreme example of a controlled experiment on vision perception was conducted with cats showing that environmental context can render a perfectly healthy cat to be essentially blind (Hubel and Wiesel, 1964). By developing more humane experiments where the design of space is highly controlled, researchers can likely assess how changes in the environment effect these animals much easier then with humans. Likewise, there is plenty of literature where the relationship between human and environment was studied, that could warrant parallels with animals. For instance, views of nature or exposure of nature has shown to reduce stress in humans (Ulrich et al., 1991) and improve well-being

(Kaplan, 2001). Thus, it is likewise worth considering how natural environments might also influence animals under various contexts, particularly within confinement.

Finally, this section shall briefly outline some domain specific areas and context in animal sensors that could benefit from an explicit focus on affective computing as a framework. The pet industry will obviously benefit from such an approach and has already pursued an affective approach in an ad hoc and limited way. However, the authors believe the potential in an affective computing approach is largely untapped in the animal health and industry. Especially in the detection of animal affect for comfort in the prevention of disease. As this chapter showed in a previous section, researchers live in a time when sensor devices are becoming ubiquitous. It is not a huge leap, to envision a time when animal telemetry devices that record the number of steps/day, heart rate, body temperature, drinking events, eating events, resting events are also ubiquitous. A veterinarian uses standard clinical observations and experience to estimate the probability that an animal has a specific disease. Common observations include temperature, stance, walking, drinking behavior, eating behavior, resting, and lethargy. The hypothesis is that sensor fusion of sensors such as pedometers, temperature, eating detection, drinking detection, and resting detection can provide machine learned models for specific disease probability estimation. If disease probability can be discerned earlier using an effective computing approach, perhaps even before an animal is contagious, the savings to the industry, the security to the food supply and cascading effects become quite significant. The authors believe this is the largest potential benefit for an affective computing approach to animal science and sensors.

Employing an affective computing framework for the explicit detection of animal affect using mobile wearable sensors is relatively nascent, so it follows that this section will now briefly discuss a potential methodology for the detection of animal arousal and affect using sensors. First, the remaining section will briefly discuss data provenance and potential data sources. Second, it will then discuss the nature of the data that will be collected. Finally, the authors propose a design for data processing and analysis.

Data provenance is essentially data likely to have been collected by the animal sensors. For example, the rudiment bolus sensor by SmartsStock will provide several parameters such as cow id, temperature, previous observations recorded, and so on. The pedometer sensor could, depending on the manufacturer either give accelerometer data in 3 dimensional axes which is continuous or process the data and provide discrete step data. Location data, using GPS coordinates or radio tags, provides an opportunity to assess movements resulting from individual agency or social influence that can help us understand how environmental or social influence trigger arousal responses. Therefore, it is quite reasonable to assert that animal sensors could be used to support similar conclusions researchers find in human affect regarding the influence of built-environment (Parsons et al., 1998; Parker 2016).

The design for data processing and analysis is informed and inspired by our work for designing affective computing experiments (Yates, Brent, & Hsu, 2017). The nature of data produced by sensors can either be discrete or continuous. The nature of the data being collected can influence, in a large part, which machine learning approach is viable in detecting and prediction for affect in the animal.

An explicit affective computing centric approach for animals via sensors indicates the need for a standardized approach for the detection of core affect such as comfort, pleasure, and so on. First, the animal will be provided an external (or invasive) wearable which records their biometric data. Second, data fusion will be necessary to combine the input channels of the sensors it is wearing. For example, heart rate (HR), temperature (temp), and movement (accelerometer). Additional information may be recorded, including GPS, audio, or remote visual information. Third, it may be necessary for a trained individual to annotate and label this data before data processing in order to assist with a machine learning approach. Fourth, the data processing will occur which prepares the data for analysis. The last step depends on the affective goal of the experiment and whether the prediction target is ordinal or continuous. For example, if ordinal then perhaps linear mixed models, logistic, or support vector machines could be applied to the analysis. Conversely, if continuous, then general linear methods or time series techniques be applied. The result will be a model for affect detection given animal sensors.

CONCLUSION

The pursuit of core animal and human affect using computing affecting via animal sensors is the central theme of this chapter. While the authors believe there are certainly challenges in an interdisciplinary approach to fuse together affective computing and animal sensors, the authors believe there are rich untapped potential and opportunities. Industry seems poised to take advantage of trends in both affective computing and animal sensors. It seems natural to fuse both approaches together and deliberately and consciously pursue animal affect via animal sensors under an affective computing framework. The authors believe that our discussion of affective computing has shown that while the field is young and nascent, the progress has been rapid and the detection of affect continues to evolve and grow more sophisticated and powerful. Likewise, the authors believe the similar rapid and growing sophistication of animal sensors presents similar opportunities. In addition, focusing on shared core affect between animal and humans will help address many of the current challenges encountered when pursuing affective computing studies. Most importantly, the authors believe

Figure 6. Spatial explicit classification model for animal affective computing
Adapted from Yates, Brent, and Hsu 2017.

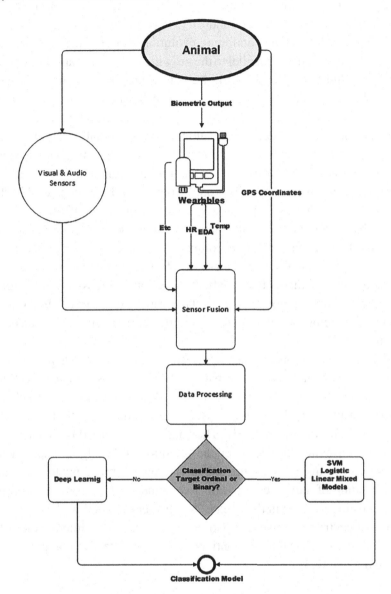

and enthusiastically embrace the opportunities of applying the growing powers and abilities of machine learning to affective computing with animal sensors.

The authors believe an explicit focus on affective computing via animal sensors has several potential advantages and open questions. Researchers can take advantage of both trends of affective computing applications and animal sensors in industry by fusing their approach. In addition, the authors want to emphasize our belief that opportunities for research and commercialization from this approach are largely untapped in the animal health and industry. It is an open question how such an approach could influence the continuing evolution and growing sophistication of animal sensors, especially in parallel to the growing powers and abilities of machine learning. The potential for teaching machines to recognize core and basic affect for discerning disease probability would be dramatic. There is ample room to grow in the approach for affect and arousal detection in animals. The authors believe our approach for affect, arousal detection could both be unimodal, and multimodal, but the emphasis on biometric data collection means researchers have barely scratched the surface. The authors believe there is a lot of room to grow in animal sensors that focus on audio and visual data collection. While beyond the scope of this chapter, the authors believe that focusing on animal affect under an affective computing framework has potential to be informative pursuing some of the loftier goals of artificial intelligence and affective computing. The proposed approach is also likely to reduce several of the challenges of affective computing while completely sidestepping some of them such as affect detection in cross-cultural settings. In addition, the authors believe that researchers can further benefit by considering how using animal emotions might provide more understanding to human emotions such as stress and etc.

The deliberate and explicit fusion of affective computing and animal sensors for the detection of core animal affect is natural, advantageous, and beneficial to both research and industry. The authors conclude by acknowledging one of the challenges in writing on an interdisciplinary topic was adequately addressing the broad collection of work, ideas, and themes properly. The authors hope that this chapter has been successful in conveying this. However, the authors acknowledge our approach is not an exhaustive coverage of ideas, but intended to paint a broad picture. The authors believe the future of animal centric affective computing via sensors is bright and that the work already done in this area is just the very beginning. The dawn for the era of animal centric affective computing is upon us. The authors anticipate rich new developments as capabilities continue to grow in affective computing, animal sensors, and machine learning.

ACKNOWLEDGMENT

The authors would like to thank the Bill Ardrey of SmartStock for his gracious assistance in providing materials for use in this chapter. The authors would like to thank Taylor Whitaker for her assistance in providing a picture of herself wearing the Empatica E4 sensor.

REFERENCES

Bickmore, T., Gruber, A., & Picard, R. (2005). Establishing the computer–patient working alliance in automated health behavior change interventions. *Patient Education and Counseling*, *59*(1), 21–30. doi:10.1016/j.pec.2004.09.008 PMID:16198215

Bickmore, T. W., & Picard, R. W. (2004, April). Towards caring machines. In CHI'04 extended abstracts on Human factors in computing systems (pp. 1489-1492). ACM. doi:10.1145/985921.986097

Bonabeau, E., Dorigo, M., & Theraulaz, G. (1999). *Swarm intelligence: from natural to artificial systems (No. 1)*. Oxford University Press.

Bosch, N., D'Mello, S., Baker, R., Ocumpaugh, J., Shute, V., Ventura, M., & Zhao, W. (2015, March). Automatic detection of learning-centered affective states in the wild. In *Proceedings of the 20th international conference on intelligent user interfaces* (pp. 379-388). ACM.

Brehme, U., Stollberg, U., Holz, R., & Schleusener, T. (2008). ALT pedometer—New sensor-aided measurement system for improvement in oestrus detection. *Computers and Electronics in Agriculture*, *62*(1), 73–80. doi:10.1016/j.compag.2007.08.014

Clarke, A. C. (2016). *2001: a space odyssey*. Penguin.

Connellan, K., Gaardboe, M., Riggs, D., Due, C., Reinschmidt, A., & Mustillo, L. (2013). Stressed spaces: Mental health and architecture. *HERD: Health Environments Research & Design Journal*, *6*(4), 127–168. doi:10.1177/193758671300600408 PMID:24089185

Constine, J. (2017). Facebook Rolls Out AI to Detect Suicidal Posts before They're Reported. *Tech Crunch*. Retrieved from https://techcrunch.com/2017/11/27/facebook-ai-suicide-prevention/

Cooke, B., Jiang, H., & Heerman, K. (2017). *Outlook for U.S. Agricultural Trade: August 2017*. United States Department of Agriculture Economic Research Service. Retrieved from https://www.ers.usda.gov/topics/international-markets-trade/us-agricultural-trade/outlook-for-us-agricultural-trade/

Crevier, D. (1993). *AI: The tumultuous history of the search for artificial intelligence*. Basic Books.

Cummins, N., Hantke, S., Schnieder, S., Krajewski, J., & Schuller, B. (2017). Classifying the context and emotion of dog barks: a comparison of acoustic feature representations. *Compare: A Journal of Comparative Education*, *40*(05), 37.

Dictionary, O. E. (2017). *Oxford English dictionary online*. Retrieved from https://en.oxforddictionaries.com/

Fernandez, R., & Picard, R. W. (2003). Modeling drivers' speech under stress. *Speech Communication*, *40*(1), 145–159. doi:10.1016/S0167-6393(02)00080-8

Fletcher, R. R., Poh, M. Z., & Eydgahi, H. (2010, August). Wearable sensors: opportunities and challenges for low-cost health care. In *Engineering in Medicine and Biology Society (EMBC), 2010 Annual International Conference of the IEEE* (pp. 1763-1766). IEEE.

Fraser, D., & Duncan, I. J. (1998). *'Pleasures', 'pains' and animal welfare: toward a natural history of affect*. Academic Press.

Garbarino, M., Lai, M., Bender, D., Picard, R. W., & Tognetti, S. (2014, November). Empatica E3—A wearable wireless multi-sensor device for real-time computerized biofeedback and data acquisition. In *Wireless Mobile Communication and Healthcare (Mobihealth), 2014 EAI 4th International Conference on* (pp. 39-42). IEEE.

Godsk, T., & Kjærgaard, M. B. (2011, August). High classification rates for continuous cow activity recognition using low-cost GPS positioning sensors and standard machine learning techniques. In *Industrial Conference on Data Mining* (pp. 174-188). Springer.

Goertz, C. E., Shuert, C. R., Mellish, J. A. E., Skinner, J. P., Woodie, K., Walker, K. A., ... Berngartt, R. K. (2017). Best practice recommendations for the use of fully implanted telemetry devices in pinnipeds. *Animal Biotelemetry*, *5*(1), 13. doi:10.118640317-017-0128-9

Healey, J., & Picard, R. (1998, May). Digital processing of affective signals. In *Acoustics, Speech and Signal Processing, 1998*. Proceedings of the 1998 IEEE International Conference on (Vol. 6, pp. 3749-3752). IEEE.

Healey, J. A., & Picard, R. W. (2005). Detecting stress during real-world driving tasks using physiological sensors. *IEEE Transactions on Intelligent Transportation Systems*, *6*(2), 156–166. doi:10.1109/TITS.2005.848368

Helwatkar, A., Riordan, D., & Walsh, J. (2014, September). Sensor technology for animal health monitoring. *International Conference on Sensing Technology, ICST*.

Hendler, J. (2008). Avoiding another AI winter. *IEEE Intelligent Systems*, *23*(2), 2–4. doi:10.1109/MIS.2008.20

Hernandez, J., McDuff, D., Fletcher, R., & Picard, R. W. (2013, March). Inside-out: Reflecting on your inner state. In *Pervasive Computing and Communications Workshops (PERCOM Workshops), 2013 IEEE International Conference on* (pp. 324-327). IEEE.

Hernandez Rivera, J. (2015). *Towards wearable stress measurement* (Doctoral dissertation). Massachusetts Institute of Technology.

Hofstadter, D. (1980). *Gödel, Escher, Bach: An Eternal Golden Braid*. New York: Basic Books.

Hoque, M. E. (2008, October). Analysis of speech properties of neurotypicals and individuals diagnosed with autism and down. In *Proceedings of the 10th international ACM SIGACCESS conference on Computers and accessibility* (pp. 311-312). ACM.

Hubel, D. H., & Wiesel, T. N. (1964). Effects of monocular deprivation in kittens. *Naunyn-Schmiedeberg's Archives of Pharmacology, 248*(6), 492–497. doi:10.1007/BF00348878 PMID:14316385

Hudson, C. D., Bradley, A. J., Breen, J. E., & Green, M. J. (2012). Associations between udder health and reproductive performance in United Kingdom dairy cows. *Journal of Dairy Science, 95*(7), 3683–3697. doi:10.3168/jds.2011-4629 PMID:22720926

Ipema, A. H., Goense, D., Hogewerf, P. H., Houwers, H. W. J., & Van Roest, H. (2008). Pilot study to monitor body temperature of dairy cows with a rumen bolus. *Computers and Electronics in Agriculture, 64*(1), 49–52. doi:10.1016/j.compag.2008.05.009

Kaplan, R. (2001). The nature of the view from home: Psychological benefits. *Environment and Behavior, 33*(4), 507–542. doi:10.1177/00139160121973115

Kapoor, A., Burleson, W., & Picard, R. W. (2007). Automatic prediction of frustration. *International Journal of Human-Computer Studies, 65*(8), 724–736. doi:10.1016/j.ijhcs.2007.02.003

Kapoor, A., & Picard, R. W. (2002, May). Real-time, fully automatic upper facial feature tracking. In *Automatic Face and Gesture Recognition, 2002. Proceedings. Fifth IEEE International Conference on* (pp. 10-15). IEEE.

Kapoor, A., & Picard, R. W. (2005, November). Multimodal affect recognition in learning environments. In *Proceedings of the 13th annual ACM international conference on Multimedia* (pp. 677-682). ACM. 10.1145/1101149.1101300

Khandaker, M. (2009). Designing affective video games to support the social-emotional development of teenagers with autism spectrum disorders. *Annual Review of Cybertherapy and Telemedicine, 7*, 37–39. PMID:19592726

Klein, S. (2017). The World of Big Data and IoT. In *IoT Solutions in Microsoft's Azure IoT Suite* (pp. 3–13). Apress. doi:10.1007/978-1-4842-2143-3_1

Knight, W. (2016). AI Winter Isn't Coming. *MIT Technology Review*. Retrieved from https://www.technologyreview.com/s/603062/ai-winter-isnt-coming/

Koyama, K., Koyama, T., Matsui, Y., Sugimoto, M., Kusakari, N., Osaka, I., & Nagano, M. (2017). Characteristics of dairy cows with a pronounced reduction in first milk yield after estrus onset. *The Japanese Journal of Veterinary Research, 65*(2), 55–63.

Kringelbach, M. L., & Berridge, K. C. (2017). The affective core of emotion: Linking pleasure, subjective well-being, and optimal metastability in the brain. *Emotion Review*. PMID:28943891

Kruijff, E., Marquardt, A., Trepkowski, C., Schild, J., & Hinkenjann, A. (2017). Designed emotions: Challenges and potential methodologies for improving multisensory cues to enhance user engagement in immersive systems. *The Visual Computer, 33*(4), 471–488. doi:10.100700371-016-1294-0

Kumar, S., & Ng, B. (2001). Crowding and violence on psychiatric wards: Explanatory models. *Canadian Journal of Psychiatry, 46*(5), 433–437. doi:10.1177/070674370104600509 PMID:11441783

Larson, K., & Picard, R. (2005, February). *The aesthetics of reading*. In Appears in Human-Computer Interaction Consortium Conference, Snow Mountain Ranch, Fraser, CO.

Le Ngo, A. C., See, J., & Phan, C. W. R. (2017). Sparsity in Dynamics of Spontaneous Subtle Emotion: Analysis & Application. *IEEE Transactions on Affective Computing*.

LeCun, Y., Bengio, Y., & Hinton, G. (2015). Deep learning. *Nature, 521*(7553), 436–444. doi:10.1038/nature14539 PMID:26017442

Løvendahl, P., & Chagunda, M. G. G. (2010). On the use of physical activity monitoring for estrus detection in dairy cows. *Journal of Dairy Science, 93*(1), 249–259. doi:10.3168/jds.2008-1721 PMID:20059923

Mader, T. L. (2003). Environmental stress in confined beef cattle. *Journal of Animal Science, 81*(14_suppl_2), E110-E119.

Martınez-Miranda, J., & Aldea, A. (2005). Emotions in human and artificial intelligence. *Computers in Human Behavior*, *21*(2), 323–341. doi:10.1016/j. chb.2004.02.010

Martiskainen, P., Järvinen, M., Skön, J. P., Tiirikainen, J., Kolehmainen, M., & Mononen, J. (2009). Cow behaviour pattern recognition using a three-dimensional accelerometer and support vector machines. *Applied Animal Behaviour Science*, *119*(1), 32–38. doi:10.1016/j.applanim.2009.03.005

McDuff, D., Gontarek, S., & Picard, R. (2014, August). Remote measurement of cognitive stress via heart rate variability. In *Engineering in Medicine and Biology Society (EMBC), 2014 36th Annual International Conference of the IEEE* (pp. 2957-2960). IEEE.

Minsky, M. L. (1967). *Computation: finite and infinite machines*. Prentice-Hall, Inc.

Mitchell, R. S., Sherlock, R. A., & Smith, L. A. (1996). An investigation into the use of machine learning for determining oestrus in cows. *Computers and Electronics in Agriculture*, *15*(3), 195–213. doi:10.1016/0168-1699(96)00016-6

Monma, H. (2011). FGCP/A5: PaaS Service Using Windows Azure Platform. *Fujitsu Scientific and Technical Journal*, *47*(4), 443–450.

Mottram, T., Lowe, J., McGowan, M., & Phillips, N. (2008). A wireless telemetric method of monitoring clinical acidosis in dairy cows. *Computers and Electronics in Agriculture, 64*(1), 45-48.

Nasirahmadi, A., Edwards, S. A., & Sturm, B. (2017). Implementation of machine vision for detecting behaviour of cattle and pigs. *Livestock Science*, *202*, 25–38. doi:10.1016/j.livsci.2017.05.014

Page, R. M. (1962). Man-Machine Coupling-2012 AD. *Proceedings of the IRE*, *50*(5), 613–614. doi:10.1109/JRPROC.1962.288048

Panksepp, J. (2004). *Affective neuroscience: The foundations of human and animal emotions*. Oxford University Press.

Panksepp, J. (2011). Cross-species affective neuroscience decoding of the primal affective experiences of humans and related animals. *PLoS One*, *6*(9), e21236. doi:10.1371/journal.pone.0021236 PMID:21915252

Panksepp, J. (2011). The basic emotional circuits of mammalian brains: Do animals have affective lives? *Neuroscience and Biobehavioral Reviews*, *35*(9), 1791–1804. doi:10.1016/j.neubiorev.2011.08.003 PMID:21872619

Parker, R. (2016). *Your Environment And You: Investigating Stress Triggers and Characteristics of the Built Environment.* Kansas State University.

Parkinson, H. (2015). Happy? Sad? Forget age, Microsoft can now guess your emotions. *The Guardian.* Retrieved from https://www.theguardian.com/technology/2015/nov/11/microsoft-guess-your-emotions-facial-recognition-software

Parsons, R., Tassinary, L. G., Ulrich, R. S., Hebl, M. R., & Grossman-Alexander, M. (1998). The view from the road: Implications for stress recovery and immunization. *Journal of Environmental Psychology, 18*(2), 113–140. doi:10.1006/jevp.1998.0086

Picard, R. W. (1995). *Affective computing.* Academic Press.

Picard, R. W. (2001, June). Building HAL: Computers that sense, recognize, and respond to human emotion. In Photonics West 2001-Electronic Imaging (pp. 518-523). International Society for Optics and Photonics.

Picard, R. W. (2002). Affective medicine: Technology with emotional intelligence. *Studies in Health Technology and Informatics*, 69–84. PMID:12026139

Picard, R. W. (2003). Affective computing: Challenges. *International Journal of Human-Computer Studies, 59*(1), 55–64. doi:10.1016/S1071-5819(03)00052-1

Picard, R. W., & Du, C. (2002). Monitoring stress and heart health with a phone and wearable computer. *Motorola Offspring Journal, 1*, 14–22.

Picard, R. W., & Healey, J. (1997, October). Affective wearables. In *Wearable Computers, 1997. Digest of Papers., First International Symposium on* (pp. 90-97). IEEE.

Picard, R. W., Papert, S., Bender, W., Blumberg, B., Breazeal, C., Cavallo, D., ... Strohecker, C. (2004). Affective learning—a manifesto. *BT Technology Journal, 22*(4), 253–269. doi:10.1023/B:BTTJ.0000047603.37042.33

Picard, R. W., & Picard, R. (1997). *Affective computing* (Vol. 252). Cambridge, MA: MIT Press.

Picard, R. W., & Rosalind, W. (2000). Toward agents that recognize emotion. *VIVEK-BOMBAY, 13*(1), 3–13.

Picard, R. W., Vyzas, E., & Healey, J. (2001). Toward machine emotional intelligence: Analysis of affective physiological state. *IEEE Transactions on Pattern Analysis and Machine Intelligence, 23*(10), 1175–1191. doi:10.1109/34.954607

Poh, M. Z., Kim, K., Goessling, A. D., Swenson, N. C., & Picard, R. W. (2009, September). Heartphones: Sensor earphones and mobile application for non-obtrusive health monitoring. In *Wearable Computers, 2009. ISWC'09. International Symposium on* (pp. 153-154). IEEE.

Poh, M. Z., McDuff, D. J., & Picard, R. W. (2011). Advancements in noncontact, multiparameter physiological measurements using a webcam. *IEEE Transactions on Biomedical Engineering*, *58*(1), 7–11. doi:10.1109/TBME.2010.2086456 PMID:20952328

Poh, M. Z., Swenson, N. C., & Picard, R. W. (2010). A wearable sensor for unobtrusive, long-term assessment of electrodermal activity. *IEEE Transactions on Biomedical Engineering*, *57*(5), 1243–1252. doi:10.1109/TBME.2009.2038487 PMID:20172811

Poria, S., Cambria, E., Bajpai, R., & Hussain, A. (2017). A review of affective computing: From unimodal analysis to multimodal fusion. *Information Fusion*, *37*, 98–125. doi:10.1016/j.inffus.2017.02.003

Reynolds, C., & Picard, R. (2004, April). Affective sensors, privacy, and ethical contracts. In CHI'04 Extended Abstracts on Human Factors in Computing Systems (pp. 1103-1106). ACM. doi:10.1145/985921.985999

Reynolds, C. J. (2005). *Adversarial uses of affective computing and ethical implications* (Doctoral dissertation). Massachusetts Institute of Technology.

Riseberg, J., Klein, J., Fernandez, R., & Picard, R. W. (1998, April). Frustrating the user on purpose: Using biosignals in a pilot study to detect the user's emotional state. In CHI 98 Cconference Summary on Human Factors in Computing Systems (pp. 227-228). ACM.

Rohrer, B. (2012, March). BECCA: Reintegrating AI for Natural World Interaction. *AAAI Spring Symposium: Designing Intelligent Robots.*

Rosenblatt, F. (1958). The perceptron: A probabilistic model for information storage and organization in the brain. *Psychological Review*, *65*(6), 386–408. doi:10.1037/h0042519 PMID:13602029

Russell, J., (2017, October). *A Dimensional Representation of Emotion.* Speech presented at ACII 2017, San Antonio, TX.

Russell, S., Norvig, P., & Intelligence, A. (1995). A modern approach. *Artificial Intelligence. Prentice-Hall. Egnlewood Cliffs*, *25*, 27.

Rutten, C. J., Velthuis, A. G. J., Steeneveld, W., & Hogeveen, H. (2013). Invited review: Sensors to support health management on dairy farms. *Journal of Dairy Science, 96*(4), 1928–1952. doi:10.3168/jds.2012-6107 PMID:23462176

Sano, A., & Picard, R. W. (2013, September). Stress recognition using wearable sensors and mobile phones. In *Affective Computing and Intelligent Interaction (ACII), 2013 Humaine Association Conference on* (pp. 671-676). IEEE.

Sano, A., Picard, R. W., & Stickgold, R. (2014). Quantitative analysis of wrist electrodermal activity during sleep. *International Journal of Psychophysiology, 94*(3), 382–389. doi:10.1016/j.ijpsycho.2014.09.011 PMID:25286449

Scheirer, J., Fernandez, R., & Picard, R. W. (1999, May). Expression glasses: a wearable device for facial expression recognition. In CHI'99 Extended Abstracts on Human Factors in Computing Systems (pp. 262-263). ACM.

Schuller, B. W. (2017). IEEE Transactions on Affective ComputingźChallenges and Chances. *IEEE Transactions on Affective Computing, 8*(1), 1–2. doi:10.1109/TAFFC.2017.2662858

Schutz, M. M., & Bewley, J. M. (2009). Implications of changes in core body temperature. In *Tri-State Dairy Nutrition Conference* (pp. 39-50). Academic Press.

Shannon, C. E. (1953). Computers and automata. *Proceedings of the IRE, 41*(10), 1234–1241. doi:10.1109/JRPROC.1953.274273

Sharkey, A., & Sharkey, N. (2011). Children, the elderly, and interactive robots. *IEEE Robotics & Automation Magazine, 18*(1), 32–38. doi:10.1109/MRA.2010.940151

Shead, S. (2017). Facebook's AI boss: 'In terms of general intelligence, we're not even close to a rat'. *Business Insider*. Retrieved from http://www.businessinsider.com/facebooks-ai-boss-in-terms-of-general-intelligence-were-not-even-close-to-a-rat-2017-10

Shults, M. C., Rhodes, R. K., Updike, S. J., Gilligan, B. J., & Reining, W. N. (1994). A telemetry-instrumentation system for monitoring multiple subcutaneously implanted glucose sensors. *IEEE Transactions on Biomedical Engineering, 41*(10), 937–942. doi:10.1109/10.324525 PMID:7959800

Sloman, A. (1999). Review of affective computing. *AI Magazine, 20*(1), 127.

Soleymani, M., Garcia, D., Jou, B., Schuller, B., Chang, S. F., & Pantic, M. (2017). A survey of multimodal sentiment analysis. *Image and Vision Computing, 65*, 3–14. doi:10.1016/j.imavis.2017.08.003

Strauss, M., Reynolds, C., Hughes, S., Park, K., McDarby, G., & Picard, R. (2005). The handwave bluetooth skin conductance sensor. *Affective computing and intelligent interaction*, 699-706.

Tao, J., & Tan, T. (2005, October). Affective computing: A review. In *International Conference on Affective computing and intelligent interaction* (pp. 981-995). Springer. 10.1007/11573548_125

Ulrich, R. S., Simons, R. F., Losito, B. D., Fiorito, E., Miles, M. A., & Zelson, M. (1991). Stress recovery during exposure to natural and urban environments. *Journal of Environmental Psychology, 11*(3), 201–230. doi:10.1016/S0272-4944(05)80184-7

Vyzas, E., & Picard, R. W. (1999, May). Offline and online recognition of emotion expression from physiological data. In *Workshop on Emotion-Based Agent Architectures, Third International Conf. on Autonomous Agents,(Seattle, WA)* (pp. 135-142). Academic Press.

Waldren, S. E., Agresta, T., & Wilkes, T. (2017). Technology Tools and Trends for Better Patient Care: Beyond the EHR. *Family Practice Management, 24*(5), 28–32. PMID:28925624

Walker, K. A., Trites, A. W., Haulena, M., & Weary, D. M. (2011). A review of the effects of different marking and tagging techniques on marine mammals. *Wildlife Research, 39*(1), 15–30. doi:10.1071/WR10177

Ward, R. D., & Marsden, P. H. (2004). Affective computing: Problems, reactions and intentions. *Interacting with Computers, 16*(4), 707–713. doi:10.1016/j.intcom.2004.06.002

Wren, C. R., & Reynolds, C. J. (2004). Minimalism in ubiquitous interface design. *Personal and Ubiquitous Computing, 8*(5), 370–373. doi:10.100700779-004-0299-2

Yates, H., Chamberlain, B., & Hsu, W. (2017). *Spatially Explicit Classification Model for Affective Computing in Built Environments*. Paper presented at the Design for Affective Intelligence Workshop at ACII2017, San Antonio, TX.

Yates, H., Chamberlain, B., Norman, G., & Hsu, W. H. (2017, September). Arousal Detection for Biometric Data in Built Environments using Machine Learning. In *IJCAI 2017 Workshop on Artificial Intelligence in Affective Computing* (pp. 58-72).

Chapter 4
Neuro–Fuzzy Models and Applications

Sushruta Mishra
KIIT University, India

Soumya Sahoo
C. V. Raman College of Engineering, India

Brojo Kishore Mishra
C. V. Raman College of Engineering, India

ABSTRACT

The modern techniques of artificial intelligence have found application in almost all the fields of human knowledge. Among them, two important techniques of artificial intelligence, fuzzy systems (FS) and artificial neural networks (ANNs), have found many applications in various fields such as production, control systems, diagnostic, supervision, etc. They evolved and improved throughout the years to adapt arising needs and technological advancements. However, a great emphasis is given in the engineering field. The techniques of artificial intelligence based on fuzzy logic and neural networks are frequently applied together for solving engineering problems where the classic techniques do not supply an easy and accurate solution. Separately, each one of these techniques possesses advantages and disadvantages that, when mixed together, provide better results than the ones achieved with the use of each isolated technique. As ANNs and fuzzy systems have often been applied together, the concept of a fusion between them started to take shape. Neuro-fuzzy systems were born which utilize the advantages of both techniques. Such systems show two distinct ways of behavior. In a first phase, called learning phase, it behaves like neural networks that learn internal parameters off-line. Later, in the execution phase, it behaves like a fuzzy logic system. A neuro-fuzzy system is a fuzzy system that uses a

DOI: 10.4018/978-1-5225-5793-7.ch004

learning algorithm derived from or inspired by neural network theory to determine its parameters (fuzzy sets and fuzzy rules) by processing data samples. Neural networks and fuzzy systems can be combined to join its advantages and to cure its individual illness. Neural networks introduce its computational characteristics of learning in the fuzzy systems and receive from them the interpretation and clarity of systems representation. Thus, the disadvantages of the fuzzy systems are compensated by the capacities of the neural networks. These techniques are complementary, which justifies its use together. This chapter deals with an analysis of neuro-fuzzy systems. Benefits of these systems are studied with its limitations too. Comparative analyses of various categories of neuro-fuzzy systems are discussed in detail. Apart from these, real-time applications of such systems are also presented.

INTRODUCTION

The modern techniques of artificial intelligence have found application in almost all the fields of the human knowledge. However, a great emphasis is given to the accurate sciences areas; perhaps the biggest expression of the success of these techniques is in engineering field. These two techniques neural networks and fuzzy logic are many times applied together for solving engineering problems where the classic techniques do not supply an easy and accurate solution. The Neuro fuzzy term was born by the fusing of these two techniques. As each researcher combines these two tools in different way, then, some confusion was created on the exact meaning of this term. Still there is no absolute consensus but in general, the Neuro fuzzy term means a type of system characterized for a similar structure of a fuzzy controller where the fuzzy sets and rules are adjusted using neural networks tuning techniques in an iterative way with data vectors (input and output system data). Such systems show two distinct ways of behavior. In a first phase, called learning phase, it behaves like neural networks that learns its internal parameters off-line. Later, in the execution phase, it behaves like a fuzzy logic system. Separately, each one of these techniques possess advantages and disadvantages that, when mixed together, theirs cooperage provides better results than the ones achieved with the use of each isolated technique.

BACKGROUND

Fuzzy systems propose a mathematic calculus to translate the subjective human knowledge of the real processes. This is a way to manipulate practical knowledge with some level of uncertainty. The fuzzy sets theory was initiated by the authors

(Tano et al., 1996) The behavior of such systems is described through a set of fuzzy rules, like:IF <Premise> THEN <Consequent> that uses linguistics variables with symbolic terms. Each term represents a fuzzy set. The terms of the input space (typically 5-7 for each linguistic variable) compose the fuzzy partition. The fuzzy inference mechanism consists of three stages: in the first stage, the values of the numerical inputs are mapped by a function according to a degree of compatibility of the respective fuzzy sets; this operation can be called fuzzification. In the second stage, the fuzzy system processes the rules in accordance with the firing strengths of the inputs. In the third stage, the resultant fuzzy values are transformed again into numerical values; this operation can be called defuzzyfication. Essentially, this procedure makes possible the use fuzzy categories in representation of words and abstracts ideas of the human beings in the description of the decision taking procedure.

Advantages of Fuzzy Systems Are:

- Capacity to represent inherent uncertainties of the human knowledge with linguistic variables.
- Simple interaction of the expert of the domain with the engineer designer of the system.
- Easy interpretation of the results, because of the natural rules representation.
- Easy extension of the base of knowledge through the addition of new rules.
- Robustness in relation of the possible disturbances in the system.

Disadvantages of Fuzzy Systems Are:

- Incapable to generalize, or either, it only answers to what is written in its rule base.
- Not robust in relation the topological changes of the system, such changes would demand alterations in the rule base.
- Depends on the existence of a expert to determine the inference logical rules.

NEURAL NETWORKS

The neural networks try to shape the biological functions of the human brain. This leads to the idealization of the neurons as discrete units of distributed processing. Its local or global connections inside of a net also are idealized, thus leading to the capacity of the nervous system in assimilating, learning or to foresee reactions or decisions to be taken. The main characteristic of the neural networks is the fact that these structures can learn with examples (training vectors, input and output samples of the system). The neural networks modifies its internal structure and the weights of the connections between its artificial neurons to make the mapping, with a level of acceptable error for the application, of the relation input/output that represent the behavior of the modeled system.

Advantages of the neural networks are:

- Learning capacity.
- Generalization capacity.
- Robustness in relation to disturbances.

Disadvantages of the neural networks are:

- Impossible interpretation of the functionality.
- Difficulty in determining the number of layers and number of neurons.

NEURO FUZZY SYSTEMS

Since the moment that fuzzy systems become popular in industrial application, the community perceived that the development of a fuzzy system with good performance is not an easy task. The problem of finding membership functions and appropriate rules is frequently a tiring process of attempt and error. This leads to the idea of applying learning algorithms to the fuzzy systems. The neural networks, that have efficient learning algorithms, had been presented as an alternative to automate or to support the development of tuning fuzzy systems. The majority of the first applications were in process control. Gradually, its application spread for all the areas of the knowledge like, data analysis, data classification, imperfections detection and support to decision-making, etc. Neural networks and fuzzy systems can be combined to join its advantages and to cure its individual illness. Neural networks introduce its computational characteristics of learning in the fuzzy systems and receive from them the interpretation and clarity of systems representation. Thus, the disadvantages of the fuzzy systems are compensated by the capacities of the neural networks. These techniques are complementary, which justifies its use together.

Types of Neuro-Fuzzy Systems

In general, all the combinations of techniques based on neural networks and fuzzy logic can be called Neuro-fuzzy systems. The different combinations of these techniques can be divided, in accordance with (Nauck et al., 1997), in the following classes:

1. **Cooperative Neuro-Fuzzy System:** As seen in Figure 1, in the cooperative systems there is a pre-processing phase where the neural networks mechanisms of learning determine some sub-blocks of the fuzzy system. For instance, the fuzzy sets and/or fuzzy rules (fuzzy associative memories or the use of clustering algorithms to determine the rules and fuzzy sets position). After the fuzzy sub-blocks are calculated the neural network learning methods are taken away, executing only the fuzzy system.

2. **Concurrent Neuro-Fuzzy System:** As seen in Figure 2, in the concurrent systems the neural network and the fuzzy system work continuously together. In general, the neural networks pre-processes the inputs (or pos-processes the outputs) of the fuzzy system.

3. **Hybrid Neuro-Fuzzy System:** In this category, a neural network is used to learn some parameters of the fuzzy system (parameters of the fuzzy sets, fuzzy rules and weights of the rules) of a fuzzy system in an iterative way. The majority of the researchers use the Neuro-fuzzy term to refer only hybrid Neuro-fuzzy system.

In the article (Nauck et al., 1997) definition: "A hybrid neuro-fuzzy system is a fuzzy system that uses a learning algorithm based on gradients or inspired by the neural networks theory (heuristical learning strategies) to determine its parameters (fuzzy sets and fuzzy rules) through the patterns processing (input and output)". A neuro-fuzzy system can be interpreted as a set of fuzzy rules. This system can be total created from input output data or initialized with the à priori knowledge in the same way of fuzzy rules. The resultant system by fusing fuzzy systems and neural

Figure 1. Cooperative neuro-fuzzy system

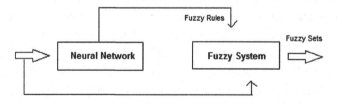

Figure 2. Concurrent Neuro-Fuzzy system

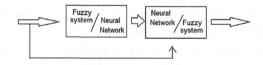

networks has as advantages of learning through patterns and the easy interpretation of its functionality. There are several different ways to develop hybrid neuro-fuzzy systems, therefore, being a recent research subject, each researcher has defined its own particular models. These models are similar in its essence, but they present basic differences.

Many types of neuro-fuzzy systems are represented by neural networks that implement logical functions. This is not necessary for the application of an learning algorithm in to a fuzzy system, however, the representation trough a neural networks are more convenient because it allows to visualize the flow of data through the system and the error signals that are used to update its parameters. The additional benefit is to allow the comparison of the different models and visualize its structural differences.

There are several neuro-fuzzy architectures like:

- Fuzzy Adaptive Learning Control Network (FALCON) (Lin & Lee, 1991);
- Adaptive Network based Fuzzy Inference System (ANFIS) (Jang, 1992)
- Generalized Approximate Reasoning based Intelligence Control (GARIC) (Berenji & Khedkar, 1992)
- Neuronal Fuzzy Controller (NEFCON) (Nauck & Kurse, 1997)
- Fuzzy Inference and Neural Network in Fuzzy Inference Software (FINEST) (Tano et al., 1996).
- Fuzzy Net (FUN) (Sulzberger et al., 1993)
- Self Constructing Neural Fuzzy Inference Network (SONFIN) (Juang et al., 1998).
- Fuzzy Neural Network (NFN) (Figueiredo & Gomide, 1999)
- Dynamic/Evolving Fuzzy Neural Network (EFuNN and dmEFuNN) (Kasabov & Qun Song, 1999).

A summarized description of the most popular neuro-fuzzy architectures is made in next section.

FALCON Architecture

The Fuzzy Adaptive Learning Control Network FALCON (Lin & Lee, 1991) is an architecture of five layers as it is shown in Figure 3. There are two linguistics nodes for each output. One is for the patterns and the other is for the real output of the FALCON. The first hidden layer is responsible for the mapping of the input variables relatively to each membership functions. The second hidden layer defines the antecedents of the rules followed by the consequents in the third hidden layer. FALCON uses an hybrid learning algorithm composed by a unsupervised learning

Figure 3. FALCON architecture

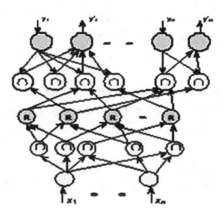

to define the initial membership functions and initial rule base and it uses a learning algorithm based on the gradient descent to optimize/adjust the final parameters of the membership functions to produce the desired output.

ANFIS Architecture

The Adaptive Network based Fuzzy Inference System ANFIS (Jang, 1992) implements a Takagi Sugeno fuzzy inference system and it has five layers. The first hidden layer is responsible for the mapping of the input variable relatively to each membership functions. The operator T-norm is applied in the second hidden layer to calculate the antecedents of the rules. The third hidden layer normalizes the rules strengths followed by the fourth hidden layer where the consequents of the rules are determined. The output layer calculates the global output as the summation of all the signals that arrive to this layer.

ANFIS uses backpropagation learning to determine the input membership functions parameters and the least mean square method to determine the consequents parameters. Each step of the iterative learning algorithm has two parts. In the first part, the input patterns are propagated and the parameters of the consequents are calculated using the iterative minimum squared method algorithm, while the parameters of the premises are considered fixed. In the second part, the input patterns are propagated again and in each iteration, the learning algorithm backpropagation is used to modify the parameters of the premises, while the consequents remain fixed.

GARIC Architecture

The Generalized Approximate Reasoning based Intelligence Control GARIC Berenji and Khedkar (1992) implements a neuro-fuzzy system using two neural networks modules, ASN (Action Selection Network) and AEN (Action State Evaluation Network). The AEN is an adaptative evaluator of ASN actions. The ASN of the GARIC is an advanced network of five layers. Figure 4 illustrates GARIC-ASN structure. The connections between the layers are not weighted. The first hidden layer stores the linguistics values of all input variables. Each input can only connect to the first layer, which represent its associated linguistics values. The second hidden layer represents the fuzzy rule nodes that determine the compatibility degree of each rule using a softmin operator. The third hidden layer represents the linguistics values of the output variables. The conclusions of each rule are calculated depending on the strength of the rules antecedents calculated in the rule nodes. GARIC uses the mean of local mean of maximum method to calculate the output of the rules. This method needs for a numerical value in the exit of each rule. Thus, the conclusions should be transformed from fuzzy values for numerical values before being accumulated in the final output value of the system. GARIC uses a mixture of gradient descending and reinforcement learning for a fine adjustment of its internal parameters.

NEFCON Architecture

The Neural Fuzzy Controller NEFCON (Nauck & Kurse, 1997) was drawn to implement a Mamdani type inference fuzzy system as illustrated in Figure 5. The connections in this architecture are weighted with fuzzy sets and rules using the same antecedents (called shared weights), which are represented by the drawn ellipses.

Figure 4. GARIC architecture

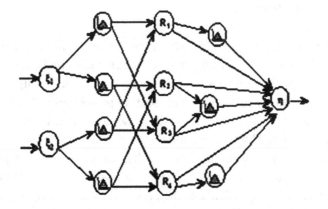

Figure 5. NEFCON architecture

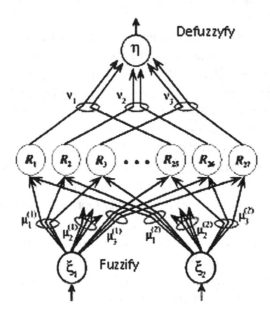

They assure the integrity of the base of rules. The input units assume the function of fuzzyfication interface, the logical interface is represented by the propagation function and the output unit is responsible for the defuzzyfication interface. The process of learning in architecture NEFCON is based in a mixture of reinforcement learning with backpropagation algorithm. This architecture can be used to learn the rule base from the beginning, if there is no à priori knowledge of the system, or to optimize an initial manually defined rule base. NEFCON has two variants NEFPROX (for function approximation) and NEFCLASS (for classification tasks) (Sulzberger, Nschichold & Vestli, 1993).

EFuNN Architecture

In Evolving Neural Fuzzy Network EFuNN (Nauck et al.,1997) all nodes are created during the learning phase. The first layer passes data to the second layer that calculates the degrees of compatibility in relation to the predefined membership functions. The third layer contains fuzzy rule nodes representing prototypes of input- output data as an association of hyper-spheres from the fuzzy input and fuzzy output spaces. Each rule node is defined by two vectors of connection weights, which are adjusted through a hybrid learning technique. The fourth layer calculates the degree to which output membership functions are matched the input data and the fifth layer carries out the defuzzyfication and calculates the numerical value for the output variable.

Dynamic Evolving Neural Fuzzy Network

This approach (Nauck et al.,1997) is a modified version of the EFuNN with the idea of not only the winning rule node's activation is propagated but a group of rule nodes that is dynamic selected for every new input vector and their activation values are used to calculate the dynamical parameters of the output function. While EFuNN implements Mamdani type fuzzy rules, dmEFuNN implements Takagi Sugeno fuzzy rules.

To get a more detail description of this architectures, beyond the specific pointed references made in this paper, a detailed survey was made by Abraham et al. (2000) where it can be found a detailed description of several well-known neuro-fuzzy architectures their respective learning algorithms.

IMPLEMENTATION OF NEURO-FUZZY SYSTEM FOR GRADING DEPRESSION

Basic purpose of our study is to develop a prototype tool which forecasts depression better than the general physicians; Neuro-fuzzy models are having wide applications in the recent times in different field with high prediction accuracies. Development of a Neuro-fuzzy-based soft-ware prototype model has been made in this study. Therefore, the main contribution of this work is the development of a Neuro-fuzzy-based prototype model for grading depression. Later, performance of the developed approach was made through two optimization methods, namely back-propagation algorithm and genetic algorithm. The model will be used by the general physicians for collecting more data. After proper validation with large set of data, it will be given for use to the psychiatrists/experts of mental health problems. Presently, the model is validated using data given by the clinicians. However, the model is evolutionary in nature and thus, it might handles uncertainty, impreciseness present (which are pre-dominant) in the system in much broader way.

Collecting depression data is very difficult. It is mainly because of the non-availability of the data. It may of different reasons which are indicated like social stigma and also majority of patients do not approach for psychiatrists. Rather they approach to general physicians and when it becomes severe they come to psychiatrists. A study is made on 312 patients reported to one of the famous mental hospital in India in the year 2004-2005. Those patients reported to the hospital for the first time and they did not take any medicines before reporting to the hospital. They are then grouped in two. First group belongs to the patients who have suicidal ideations and require immediate treatment. However, patients belonging to the second group (total

88 patients) were further analyzed towards the severity of depression. Therefore, there exist 88 data; out of which 78 has been used for training/ optimizing the Neuro-fuzzy architecture and rest 10 have been used for testing the performance of optimized architecture. Seven common symptoms (FS, LP, WL, IN, H, LA, PA) are then observed by three senior psychiatrists. Doctors are then requested to quantify the symptoms and severity of depression with which those 88 patients are suffering.

- FS-Feeling Sad
- LP-Loss of Pleasure
- WL-Weight Loss
- IN-Insomnia
- H-Hyper insomnia
- LA-Loss of Appetite
- PA-Psychomotor Agitation

Development of Neuro-Fuzzy Model

A Mamdani-type page7NN-FL model was developed by Hui et al. is used in this study for grading depression. Interested readers may go through the paper of Hui et al. (2000) for detailed discussion on the development of NN-FL system. In the present study, seven inputs representing the seven different symptoms and one output representing the depression is used. Three different grades of the inputs are considered (m, M and s). The data base of the NN-FL system is shown in Figure 6. Since there are three possibilities (i.e., mild, moderate and severe) of the load of each of seven inputs, there will be 37 = 2187 numbers of possible input combinations, for which grades of depression would vary as mild, moderate and severe, based on the symptom load. Manually constructed rules looks like:

If FS is m & LP is m & WL is m & IN is m & H is m & LA is m & PA is m THEN Depression is M.

The performance of the above NN-FL system largely depends on those rules and different weights such as V1, V2, V3, V4, V5, V6, V7 and W1. The NNFL system will have variable structure only when rule base of the FLC is optimized i.e., some rules are deleted. There exist a large number of literatures for developing an optimal NNFL system. In the present study two different approaches have been adopted. Both these two approaches have been explained below in brief.

Figure 6. The proposed NN-FL system

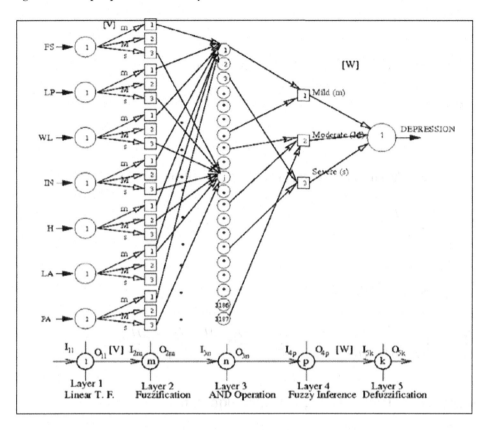

Approach 1: Tuning Using a BP Algorithm

In this approach, RB is kept unaltered during its training. Only the weights V1, V2, V3, V4, V5, V6, V7 and W1 are optimized using BP algorithm. It is usually implemented through the mini-mization of Mean Squared Error (E) in prediction, as given below.

$$E = \frac{1}{2C} \sum_{C=1}^{C} \left(T_{5_C} - O_{5_C} \right)^2$$

where T5c and O5c represent the target and model predicted outputs of cth training scenario.

The change in W1 that is DW1 is determined as

$$\Delta W_1 = -\Delta \frac{\partial E}{\partial W_1}$$

The change in W1 which is DW1 is determined as:

$$\frac{\partial E}{\partial W_1} = \frac{\partial E}{\partial O_{5_C}} \frac{\partial O_{5_C}}{\partial I_{5_C}} \frac{\partial I_{5_C}}{\partial W_1}$$

Similarly, the change in Vi that is DVi can be calculated as follows:

$$\Delta V_1 = -\Delta \frac{\partial E}{\partial V_1}$$

$\frac{\partial E}{\partial V_1}$ is calculated as

$$\frac{\partial E}{\partial V_1} = \frac{\partial E}{\partial O_{5_C}} \frac{\partial O_{5_C}}{\partial I_{5_C}} \frac{\partial I_{5_C}}{\partial O_{4P}} \frac{\partial O_{4P}}{\partial I_{4P}} \frac{\partial I_{4P}}{\partial O_{3K}} \frac{\partial O_{3K}}{\partial I_{3K}} \frac{\partial I_{3K}}{\partial O_{2M}} \frac{\partial O_{2M}}{\partial I_{2M}} \frac{\partial I_{2M}}{\partial V_i}$$

Approach 2: Tuning Using Genetic Algorithm

Approach 1 suffers from the local minima problem. Also, in Approach 1, only the weights are optimized. Therefore, in Approach 2, optimization of both DB and RB is attempted using a GA. GA with 4454-bits long string is used for this purpose. Data base is represented by first Figure 6.

Since GA is a population-based search and each population is to be assigned a Þtness, mean squared error in prediction of degree of depression is considered as the fitness for each population. GA aims at Þnding that particular solution which gives lowest error in predicting degree of depression. Also, through this approach, it is possible to reduce the number of rules.

It is very difficult for a doctor to remember a large number of rules. Some of the GA-optimized rules may be redundant (are not important) in the context of the problem. Therefore, it will be nicer to eliminate them from the rule base. However, identification of redundant rules is difficult. In this paper, we have planned to identify them using the concept of importance factor. Rules having relatively low importance factors are called as redundant and may be deleted from the rule base.

Let us suppose that nij denotes the number of times that the jth linguistic term (j = 1 for mild, j = 2 for moderate and j = 3 for severe) for the ith input condition (i.e., corresponding to different input values) is fired during training. In this way, total numbers of fired rules are represented by N. Thus, the probability of occurrence of the jth linguistic term for the ith input condition may be given as follows

$$P_{ij} = \frac{n_{ij}}{N},$$

where i=1,2,3 --------2187 and j=1,2,3.

The probability of occurrence of a rule will indicate the necessity of keeping that rule in the RB. It is then multiplied with the worth of a rule to determine the importance factor. The worth of any linguistic term for the depression follows a Gaussian distribution pattern.

Let us say that, Cj is the worth of the jth linguistic term

$$C_j = \frac{1}{2\pi^2} e^{-\frac{1}{2}\left(\frac{x_j-1}{\sqrt{2\pi}}\right)^2}$$

where xj=0.005+(j-1)W

W denotes the half of the triangular base of the data base and is different for different chromosomes of GA. The average worth over the optimization procedure is taken into account for the determination of importance factor of a rule.

The importance factor of a rule may be determined as follows:

$$I_i = \frac{1}{2}\left(P_{ij} \times C_j\right)$$

A particular rule is considered to be redundant if the importance factor of that rule is very low and can be eliminated only when non-firing does not arise. Let us understand the same through an example. If we consider, S = 0.3, P = 0.3, W1 = 0.3, I1 = 0.3, H = 0.3, A = 0.3 and P1 = 0.3. So, all the inputs can be classified as either mild (m) or moderate (M). Therefore, at a time 27 = 128 rules out of existing 2187 rules will be fired. Now, out of those 128 rules, if one such rule is absent in the rule base, there will be a non-firing of such a rule. One training/test scenario represents one input set, resulting in possibility of a maxi-mum 128 nonÞring situations. Therefore, during training with 78 training cases, there is a possibility of 128* 78 = 9984 non-firing situations, if for every case all the fired rules are absent. Therefore maximum possible non-firing situations = maximum number of generations populations * 9984 in case of FLC tuned with GA.

RESULTS AND DISCUSSIONS

Two approaches as explained above sections are used to train NN-FL system. Out of the 88 data, 78 are used for training and 10 are used for validation. During this process, all the weights are varied in between 0.1 and 0.45. In Approach 1, learning rate (g) is systematically varied in range of 0.01Ð0.5 and result is noted. shows the variation of g with MSE and best result is obtained with g = 0.05. This study is done

Figure 7. DB of the input and output variables

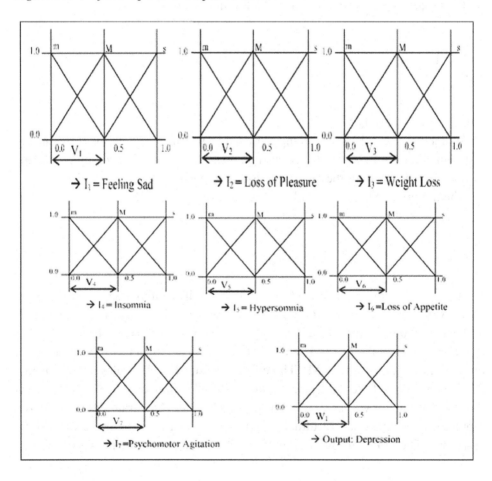

considering the number of iterations equal to 15. Lowest MSD is found to be equal to 0.0116043 (refer to. Through this process optimal weights are found to be V1 = 0.4161; V2 = 0.4469; V3 = 0.4507; V4 = 0.4469; V5 = 0.4405; V6 = 0.4232; V7 = 0.4357 and W1 = 0.3425.

An attempt has also been made to identify the redundant rules and thereafter those rules have been removed from the rule base. After removing certain number of rules which were found to have lower values of importance factor, the number of non-fired scenarios and the corresponding errors for both the training set (containing 78 data sets) and the test set (containing 10 data sets) were calculated and Zero non-fired scenarios ha e been observed with the removal of 500 rules out of 2187 during test cases and only 25 non-Pred situations happened for the training data-sets. Therefore, final result has been taken removing 500 rules from the rule-base. The reason behind removing 500 rules is (i) it does not give rise to any non-firing situations of the rule base for the test cases and (ii) no further improvement in error in prediction. It has been observed that there exist three different zones of importance factors. Moreover, nearby rules (say rule no 1819 and 1820) have same large importance fac-tors. It might have happened because of the following reasons:

1. There are only three linguistic terms representing depression (mild, moderate and severe). Therefore, there could be three different zones of the importance factor. Each zone will have some mean and a range,
2. high value of importance factor means more number of firing of such a rule during the process. Since we have considered triangular membership function distributions, which are symmetric in nature, therefore, rules having linguistic terms mild and moderate will have chance of firing at the same time.

Figure 8. MSE vs. learning rate

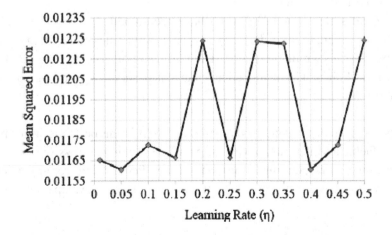

Figure 9. MSE vs. iteration number with learning rate=0.5

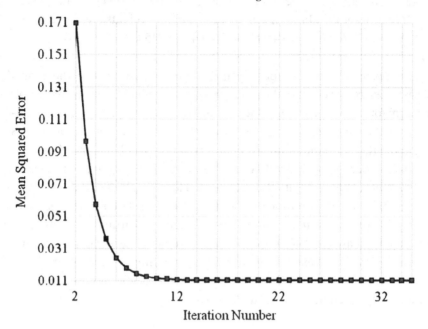

Following things have been observed in the rule base.

Type A: 835 rules out of total 2187 rules have not been modifed by the GA. It clearly indicates that those are the rules in principle and even their important factor is low, still they can never be omitted from the rule base

Type B: 1106 rules out of 2187 rules have been modified slightly i.e mild to moderate, moderate to severe, moderate to mild, severe to moderate. It indicates that there is a overlapping in decision making of doctors and they could not clearly defined the rules.

Type C: 286 rules out of total 2187 rules have been modifed largely (i.e., mild to severe and severe to mild. These are hypersensitive rules and existence in the rule base causes wrong prediction. Therefore, authors tried to eliminate those rules as a whole and have noticed no change is observed in terms of nonfringe scenarios.

It is also important to mention that rule no. 1823 is the best rule in terms of importance factor and it is also a rule in principle. Table 4 shows the predicted depression values corresponding to the ten test cases. Percentage error in prediction is found to be low leaving the tenth case. It indicates that the developed model is a

Table 1. Non-fired Scenarios Vs Number of rules removed

Total Rules Removed From 2187 Rules Present	No of Non-Firing Scenarios		Error in Prediction	
	Training Cases	Test Cases	Training Cases	Test Cases
2000	3312	440	0.012633359	0.01368848
1600	1464	188	0.012357538	0.01330318
1500	1280	160	0.012655688	0.01368225
1400	1011	133	0.012416698	0.01368225
1300	862	105	0.012416698	0.01368225
1200	615	80	0.012416698	0.01368225
1100	507	61	0.012416698	0.01368225
1000	315	37	0.012416698	0.01368225
500	25	0	0.012416698	0.01368225
105	0	0	0.012416698	0.01368225

Table 2. Some important rules

Rule No	FS	LP	WL	IN	H	LA	PA	Depression		Importance Factor
								Manual	Optimized	
1	M	m	m	m	m	m	m	m	s	1.26×10^{-9}
9	M	M	m	m	m	s	s	s	m	2.62×10^{-9}
1795	S	M	M	m	M	M	m	M	s	0.000189
1796	S	M	M	m	M	M	M	M	m	0.00019
1822	S	M	M	M	M	M		M	M	0.000208
1823	S	M	M	M	M	M	M	M	M	0.00021

good fit one. Moreover, less percentage error is seen with Approach. It may be due to the optimized rule base in the approach. Performance of the

best NN-FL system i.e., Approach 2 is com-pared with a NN-based classifier system. The NN- based system will have a three layered 7-6-1 architecture. It has been observed that the predicted depression by Approach 2 is much closer to the target depression compared to NN-based system. It may be due to the fact that logic-based controller works well in modeling depression of humans.

Table 3. Predicted depression vs targeted depression

No	FS	LP	WL	IN	H	LA	PA	TD	PD		%Error	
									App. 1	App. 2	App. 1	App. 2
1	0.7	0.8	0.5	0.5	0.5	0.6	0.7	0.59	0.6216	0.6182	-5.37	-4.78
2	0.6	0.5	0.9	0.9	0.6	0.9	0.9	0.66	0.6191	0.6161	6.18	6.65
3	0.8	0.6	0.6	0.7	0.7	0.6	0.6	0.57	0.6032	0.5902	-5.83	-3.54
4	0.7	0.4	0.6	0.8	0.6	0.8	0.4	0.60	0.6521	0.6524	-8.69	-8.73
5	0.9	0.5	0.7	0.8	0.6	0.5	0.5	0.60	0.6178	0.6028	-2.97	-0.46
6	0.9	0.6	0.7	0.7	0.8	0.6	0.6	0.68	0.6070	0.6643	10.72	5.24
7	0.6	0.4	0.7	0.6	0.6	0.5	0.8	0.72	0.6557	0.6802	8.91	5.52
8	0.5	0.9	0.4	0.6	0.8	0.7	0.8	0.87	0.6879	0.7659	19.77	11.96
9	0.5	0.6	0.6	0.6	0.9	0.6	0.7	0.70	0.6519	0.6813	6.86	2.67
10	0.6	0.4	0.6	0.5	0.6	0.6	0.6	0.60	0.6509	0.6182	34.24	37.56
Average									0.698	0.6377	0.6489	
Standard Deviation									0.1351	0.0292	0.0522	

Table 4. Predicted depression using approach 2 & NN based classifier system

Test Case No.	Target Depression	Approach 2	NN-Based System
1	0.59	0.6182	0.5263
2	0.66	0.6161	0.6070
3	0.57	0.5902	0.4047
4	0.6	0.6524	0.3870
5	0.6	0.6028	0.3348
6	0.68	0.6643	0.5010
7	0.72	0.6802	0.4019
8	0.87	0.7659	0.8133
9	0.7	0.6813	0.6110
10	0.99	0.6182	0.0741
Average	0.698	0.6489	0.4661
Standard deviation	0.1351	0.0522	0.1976

CONCLUSION

Diagnosis of depression consumes months to get diag-nosed even by an expert psychiatrist and therefore speed is out of the scope of the paper. Still, to reduce the time the paper attempts to identify the redundant rules and analysis has been made to remove them by identifying the load of individual rules on the morbidity. Finally, performance of the optimized GA-tuned NNFL system is compared with that of back-propagation based NNFL system for ten different cases. It was found that the performance of GA-tuned NNFL system is better compared to the other one. Moreover, it has been observed that GA-tuned NNFL sys-tem has outperformed to a NN-based system. Therefore, Approach 2 can be of great help to the doctors for diagnosing the patients suffering from depression.

REFERENCES

Abraham, A., & Baikunth, N. (2000). *Hybrid Intelligent Systems: A Review of a decade of Research. School of Computing and Information Technology.* Autsralia *Technical Report Series (World Health Organization), 5/2000,* 1–55.

Berenji, H. R., & Khedkar, P. (1992). Learning and Tuning Fuzzy Logic Controllers through Reinforcements. *IEEE Transactions on Neural Networks, 3*(5), 724–740. doi:10.1109/72.159061 PMID:18276471

Czogala, E. & Leski, J. (2000). *Neuro-Fuzzy Intelligent Systems, Studies in Fuzziness and Soft Computing.* Springer Verlag.

Figueiredo & Gomide. (1999). Design of Fuzzy Systems Using Neuro-Fuzzy Networks. *IEEE Transactions on Neural Networks, 10*(4), 815-827.

Jang, R. (1992). *Neuro-Fuzzy Modelling: Architectures, Analysis and Applications* (PhD Thesis). University of California, Berkeley, CA.

Juang, F. C., & Chin Lin, T. (1998). An On-Line Self Constructing Neural Fuzzy Inference Network and its applications. *IEEE Transactions on Fuzzy Systems, 6*(1), 12–32. doi:10.1109/91.660805

Kasabov, N., & Song, Q. E. (1999). *Dynamic Evolving Fuzzy Neural Networks with 'm-out-of-n' Activation Nodes for On-Line daptive Systems.* Technical Report TR99/04, Departement of Information Science, University of Otago.

Kosko, B. (1992). *A Dynamical System Approach to Machine Intelligence. In Neural Networks and Fuzzy Systems.* Englewood Cliffs, NJ: Prentice Hall.

Lin, T. C., & Lee, C. S. (1991). Neural Network Based Fuzzy Logic Control and Decision System. *IEEE Transactions on Computers*, *40*(12), 1320–1336. doi:10.1109/12.106218

Nauck, D. (1994). A Fuzzy Perceptron as a Generic Model for Neuro-Fuzzy Approaches. *Proc. Fuzzy-Systems, 2nd GI-Workshop*.

Nauck, D. (1995). Beyond Neuro-Fuzzy Systems: Perspectives and Directions. *Proc. of the Third European Congress on Intelligent Techniques and Soft Computing*.

Nauck, D., Klawon, F., & Kruse, R. (1997). *Foundations of Neuro-Fuzzy Systems*. J. Wiley & Sons.

Nauck, D., & Kurse, R. (1997). Neuro-FuzzySystems for Function Approximation. *4th International Workshop Fuzzy-Neuro Systems*.

Sulzberger, S., Nschichold, E., & Vestli, S. (1993). FUN: Optimization of Fuzzy Rule Based Systems Using Neural Networks. *Proceedings of IEEE Conference on Neural Networks*, 312-316.

Tano, S., Oyama, T., & Arnould, T. (1996). Deep Combination of Fuzzy Inference and Neural Network in Fuzzy Inference. *Fuzzy Sets and Systems*, *82*(2), 151–160. doi:10.1016/0165-0114(95)00251-0

Zadeh, L. A. (1965). Fuzzy Sets. *Information and Control*, *8*(3), 338–353. doi:10.1016/S0019-9958(65)90241-X

Chapter 5
Human Health Risk Assessment via Amalgamation of Probability and Fuzzy Numbers

Palash Dutta
Dibrugarh University, India

ABSTRACT

This chapter presents an approach to combine probability distributions with imprecise (fuzzy numbers) parameters (mean and standard deviation) as well as fuzzy numbers (FNs) of various types and shapes within the same framework. The amalgamation of probability distribution and fuzzy numbers are done by generating three algorithms. Human health risk assessment is performed through the proposed algorithms. It is found that the chapter provides an exertion to perform human health risk assessment in a specific manner that has more efficacies because of its capacity to exemplify uncertainties of risk assessment model in its own fashion. It affords assistance to scientists, environmentalists, and experts to perform human health risk assessment providing better efficiency to the output.

INTRODUCTION

Health risk estimates whether inhabitants may be exposed to harmful compounds from different sources and if so is the matter, whether those exposure intensities corroborate an undesirable health risk based on toxicological swots up. The risk evaluation process may engage investigating different pathways such as inhalation of

DOI: 10.4018/978-1-5225-5793-7.ch005

air, intake of water, intake of food, and/or dermal contact etc. The natural environment is one of the sources of radiation/exposure. However, artificial or manmade sources also play crucial rule for exposure of toxic compounds. Generally, human and biota are affected through the above-mentioned pathways.

Once perilous essences are discharged into the environment, an assessment is indeed obligatory to find out potential impact these essences may have on human health and environment. Health risk estimation is necessitated when contamination is detected above a particular threshold value. Human health risk assessment includes four basic steps viz., Hazard Identification, Dose-Response Assessment, Exposure Assessment and Risk Characterization. Hazard identification investigates either a compound has the possibility to source of impairment to humans and/or biota, or if it happens, it is necessary to find "is this compound harmful to humans?" Dose-response assessment investigates the numerical relationship between exposure and effects i.e., what amount of injury is this level of exposure likely to cause? Exposure assessment investigates how much of the compound are people being exposed to over what time period? Once Dose-response assessment and data on exposure are achieved, risk characterization can be performed by formulating a mathematical model to evaluate human being or inhabitant's risk. This can be obtained by adding the impact over all exposure pathways. One can say, what is the extra risk to human health caused by this amount of exposure to this compound? However, these basic processes of risk assessment involve uncertainties. Furthermore, a health risk is generally performed through mathematical model which is a function of components. But, it not always possible to obtain accurate values with precision of the model components. Generally, two types of uncertainties are considered in risk assessment viz., aleatory Uncertainty and epistemic Uncertainty. The former type of Uncertainty occurs because of randomness, variability, stochasticity etc. and this type of uncertainty can't be reduced. It may be tied to variations in physical and biological process and cannot be reduced with additional research or information although it may be known with greater certainty. The data like demographic data on food intake, water intake which depends on the height, body weight, socio-economic status life style and inherent variation in dietary habits faces this type of uncertainty. Variability usually has different levels. For example, daily consumption rate of meat/fish vary from person to person, vary between consecutive days, vary between seasons, vary between countries *etc.* Similarly, height, hair colour, food consumption, breathing rate etc vary. This type of uncertainty cannot be reduced by any means. However, the uncertainty can be reported with high level of confidence and can be modelled using various statistical test and tools. Later type of uncertainty occurs due to lack of knowledge, imprecision, lack of data, partial information etc. When dealing with these kinds of uncertainty one often has to rely on experts and their subjective judgements. So, it is also known as subjective uncertainty. Incertitude, ignorance, non-specificity, reducible uncertainty are other terms used for this uncertainty. Unlike variability, epistemic uncertainty can be reduced by further study such as additional experimentation, collection of additional bit of data *etc.* For example, model uncertainty belongs to this type of uncertainty this can be reduced with more understanding of the substantial situation. Sometimes parameter uncertainty belongs to epistemic uncertainty because of lack of information (Vose, 2002).

Consequently, representation of some uncertain models parameters may be of probability distributions with representations of mean and standard deviation are bell-shaped fuzzy numbers (BFNs) or simply fuzzy numbers (FNs), while representation of some other model parameters may be fuzzy numbers of different types and shapes, then it creates complicacy in the risk assessment process. On the other hand, based on accrued data/information, other components of the model can be represented by BFNs. To deal with such situation, one must has to develop a new efficient combined technique.

When BFNs are involved in the model, it becomes impractical to execute the model because of different alpha values are obtained for the same confidence interval. Thus, coexistence of probability distributions with representations of the parameters (*e.g.*, mean and standard deviation) of probability distributions are BFNs as well as FNs in health risk assessment makes the situation complex.

Therefore, the foremost objective of this article is to make an effort for amalgamations of Probability distributions where representations of parameters (mean & standard deviation) are of BFNs as well as fuzzy numbers (FNs) of various types and shapes within the same framework. For this purpose, it is proposed to generate three algorithms. Finally, health risk assessment will be carried out using the proposed algorithms.

LITERATURE REVIEW

As in health risk assessment above mentioned uncertainties co-exist. Hence, combined dissemination of above discussed theories and their application in risk assessment are addressed by various researchers for instance Probabilistic-fuzzy health risk (Guyonnet et al., 1999, 2003; Kentel & Aral, 2004; Baudrit & co-workers 2006, 2008, 2008; Anoop et al., 2008; Pedroni et al., 2012, 2013), Monte Carlo and possibilistic approach in event tree analysis (Baraldi et al., (2008), through evidence theory (Limbourg et al., 2010), hybrid fuzzy-stochastic modelling (Chen & Lee, 2010), probabilistic and Possibilistic treatment (Flage et al., 2010, 2011), Fuzzy logic and adaptive neuro-fuzzy (Karami et al., 2013), fuzzy set theory and Monte Carlo simulation (Arunraj & Mandal (2013), blueprint for evaluation of risk in 21st century (Pastoor et al., 2014), estimation of risk for Salmonella in tree nuts, in low-water activity foods and in low-moisture foods (Santillana et al., 2013, 2014, 2016), uncertainty modelling for risk assessment and risk management for safe foods (Zwietering (2015), evaluation of investment project risk in the presence of imprecision and randomness (Rębiasz et al., 2016), uncertainty modelling for evaluation of risk for container port (Alyami et al., 2016), uncertainty handling in safety instrumented systems via Monte Carlo analysis and fuzzy sets (Innalet et al., 2016), randomness and fuzzy theory for uncertainty quantification in dynamic system risk assessment (Abdo & Flaus, 2016), risk assessment of shallow groundwater contamination under irrigation and fertilization conditions (Zhang et al., 2016), health risk evaluation through evidence theory with generalized fuzzy number (Dutta, 2015), amalgamation of generalized fuzzy numbers and probability distribution in health risk assessment (Dutta, 2013; Dutta and Ali, 2015; Dutta, 2017), possibilistic-probabilistic programming approach (Liu et al., 2017), groundwater vulnerability assessment via combined fuzzy logic models (Nadiri et al., 2017), health risk assessment and SWOT analysis (Dutta, 2018). However, the coalescence of these uncertainty modelling theories in evaluation of risk is usual, but amalgamation of probability distributions with imprecise (fuzzy) parameters (i.e., mean & standard deviation) and fuzzy numbers of different shapes & shapes are not often and not an easy task.

UNCERTAINTY MODELLING TECHNIQUES

Uncertainty is an inescapable ingredient of risk assessment. Based on the nature, attribute and comprehensibility of data, information uncertainties are modelled through Probability theory, normal FS or IVFS set, P-box etc.

Probability Theory

Probability theory is commonly employed in uncertainty modelling, occurs due variability, stochasticity, randomness etc. If the nature of the model components are random and follow existing probability distributions, this theory is more appropriate which produces again probability distributions. Probability distributions are characterized by a random variable (discrete and continuous), variable in a study in which subjects are randomly selected. In uncertainty analysis most frequently the following probability distributions are employed such as normal probability distributions, lognormal normal probability distributions, triangular probability distribution, uniform probability distribution etc. (Dutta and Ali, 2011; Dutta and Hazarika, 2017).

Fuzzy Set Theory

FST is pertinent to embody epistemic uncertainty occurs because of miniature data, imprecision, lack of information, insufficiency of data etc. In such environments, classical theory of probability is not proper/appropriate to plough this type of uncertainty. FST is extensively use to plough uncertainty which absolutely allots a value in between 0 and 1 that indicates the believe/grade of membership/association of identifiable element in the set. Generally, triangular fuzzy numbers (TFNs) or trapezoidal fuzzy numbers (TrFNs) are expansively deliberated. However, in real world situations, there are some applications where BFNs may occur to plough epistemic uncertainty. It is palpable intuition that the ordinary fuzzy set is not capable to allocate precise value in between 0 and 1 (Gehrke et al., 1996) and on the other hand asking for precise vale to expert is imprudently confining. Therefore, allocating of interval degree is more empirical. In such situations interval valued fuzzy set (IVFS) move towards into existence. The concept of IVFS (ϕ − fuzzy set) was first developed by Sambuc (1975). An IVFS is defined to be a degree of association is a closed subinterval of [0,1] that consists of upper membership function (UMF) and other is lower membership function (LMF). (Zadeh, 1965; Ashraf et al., 2007; Das & Dutta, 2013) are given in this section.

Probability Box (P-Box)

In some situations, components of probability distributions i.e., mean and variance are commonly pretentious by epistemic uncertainty because of lack of experience and knowledge on the situation represented by the probability distributions. In such circumstances, p-box can be formulated which unites probability distributions and arithmetic of intervals. One can say, if the cumulative distribution functions lies on or between two monotonic curves, it produces a p-box. When the distributions are exactly precise the bounds will be coincide. A p-box has the ability to represent both types of uncertainties in the same times. In literature two types of p-boxes are obtained, viz., parametric and non-parametric p-box (Dutta and Ali, 2011; Dutta and Hazarika, 2017).

APPROACHES TO UNITE PROBABILITY DISTRIBUTIONS WITH IMPRECISE PARAMETERS AND FUZZY NUMBERS OF DIFFERENT SHAPES AND TYPES:

It is well established that co-existence of uncertainties is a common phenomenon in human health risk assessment. It is further experienced that certain model components are reasonably embodied by probability distributions while others are better represented by FNs. Then, it is essential to amalgamate the two modes of representation i.e., to lever hybrid uncertainties (*i.e.,* fuzzy and Probability distributions) for the purpose of assessing the risk, In uncertainty modelling for probability distributions the parameters are generally taken as ordinary or crisp set, but because of occurrence of uncertainty it it should be adopted as fuzzy numbers with different shapes.

Here, three algorithms have been devised to amalgamate probability distributions with fuzzy parameters of different shapes and FNs of different shapes and types.

Consider any arbitrary Mathematical model

$$M = f(P_j, B_k, T_l), j = 1, 2, 3, ..., r; k = 1, 2, 3, ..., s; l = 1, 2, 3, ..., t.$$

The model is a function of the parameters P_j and B_k and T_l. Depictions of Dj are probability distributions where parameters are BFNs or TFNs; depictions of B_k are normal or interval valued BFNs and T_l are normal or interval valued TFNs.

Algorithm-I

In this algorithm, representations of some uncertain model parameters are considered as probability distributions with fuzzy parameters. While it is considered that representations of some other uncertain model parameters are also normal FNs. First consider all those the fuzzy numbers which represent mean & standard deviation of the probability distributions. Then, evaluate alpha cut of the fuzzy numbers for alpha value 0 that gives two intervals for each probability distribution. All the combinations of the two intervals of the respective probability distributions are taken and CDFs are plotted. In this process four cdfs are obtained. Taking the envelope over the four cdfs of the probability distributions produces a p-box. Again, same alpha value is considered for other FNs and calculated alpha-cuts which produce closed intervals. Finally, assign all interval values as well as all p-boxes in the model that produces a p-box. The process is repeated for N times. Also, process is repeated for other α − cuts. Thus, a set of p-boxes is acquired, from where membership functions (MFs) are produced at different fractiles and that are obtained in the form of FNs.

Suppose of P_j are probability distributions in which parameters are considered as TFNs; representations of B_k and T_l are considered as TFNs.

The procedure is summarized as:

Step 1: Take each and every one the uncertain input components.

Step 2: Compute α-cut for α-level 0 for fuzzy parameter of each probability distribution. It will produce *r* numbers of p-boxes.

Step 3: Evaluate α-cut for the same α-level 0 for the other FNs. *s+t* numbers of closed intervals will be obtained.

Step 4: Put all the *r* numbers of p-boxes and the *(s+t)* number of intervals in the model *M*. It will produce the resultant p-box for α-level 0.

Step 5: Again, gauge α-cut for other α-level and repeat step2 to step4. The program will be terminated when the α reach the level 1.

After execution of the program, a set of p-boxes are obtained. For example, if α takes equally spaced 11 values (*i.e.,* $\alpha = [0 : 0.1 : 1]$) then a total of 11 p-boxes are obtained. 10 distinct p-boxes corresponds to α-level 0 to 0.9 and there is single common p-box (like a CDF) corresponds to 1-cut (i.e., $\alpha = 1$).

From this family of p-boxes, MFs (i.e., fuzzy numbers here) of risk are generated for different fractiles (Maxwell et. al., 1998; Kentel & Aral, 2004; Dutta & Ali, 2011).

Algorithm-II

In this algorithm, depictions of some uncertain model components are probability distributions with fuzzy (BFNs/TFNs) parameters. In addition depictions of remaining some other uncertain model components are also BFNs & TFNs. For execution of the model, it is needed to consider all FNs first and discretize the all the FNs at 0.01 (*i.e.,* the MF of FN are defined on [0.01,1]), evaluating alpha-cut of the FNs that generates crisp closed intervals. Then, calculate alpha cut of the fuzzy numbers for alpha value 0.01 that gives two intervals for each probability distribution. Then, proceed as algorithm-I. Here too, a set of p-boxes is obtained and from which for different fractiles MFs of FNs are produced. Suppose of P_j are probability distributions where some parameters are adopted as BFNs and some others are TFNs; representations of B_k are BFNs and T_l are TFNs.

The procedure is summarized by the steps below:

Step 1: Take each and every uncertain input model components. The convexity condition of the FNs are relaxed first by considering α-corresponding to α-cut defined in the interval [0.01,1].

Step 2: Adopt $\alpha = \left[0 : \dfrac{0.99}{q} 1\right], q \in \mathbb{N},$ and compute α-cut for α-level 0.01 for of each probability distribution. Then, here also r numbers of p-boxes are achieved.

Step 3: In the same way, compute α-cut for α-level 0.01 for the other FNs. It will produce $s+t$ numbers of closed intervals.

Step 4: Put all the r numbers of p-boxes and the $(s+t)$ number of intervals in the model M. It will produce the resultant p-box for α-level 0.01.

Step 5: Compute α-cut for other α-level and repeat step2 to step4. Here also the program will be terminated when the α reach the level 1.

Completion of the program produces a set of p-boxes. For example, here also if α takes equally spaced 11 values (*i.e.,* $\alpha = [0.01 : 0.099 : 1]$) a total of 11 p-boxes will be obtained. 10 distinct p-boxes corresponds to α-level 0.01 to 0.099 and while there will be one common p-box (like a CDF) corresponds to 1-cut (i.e., $\alpha=1$).

Here also, from this set of p-boxes, MFs of FNs of risk for different fractiles are generated.

Algorithm-III

In this algorithm, also depictions of some uncertain model components are adopted as probability distributions with fuzzy (TFNs/BFNs) parameters and some other uncertain model components are adopted as interval valued BENs/FNs. Here, first upper membership functions (UMF) of all input IVFNs are considered and execute algorithm-II. Then, again considered lower membership functions (LMF) of all input IVFNs and again followed algorithm-II. Here, two sets of p-boxes are obtained and from these families of p-boxes membership functions of resultant fuzzy sets at different fractiles are generated which are obtained in the form of interval valued almost BFNs.

Suppose P_j are probability distributions with fuzzy (BFNs & TFNs) parameters, B_k are interval valued BFNs and T_j are Interval valued TFNs.

The procedure is presented below:

Step 1: Take all the uncertain input components whose depictions are probability distributions and UMFs of interval valued fuzzy numbers (IVFNs). Here too, the convexity conditions of the fuzzy numbers (parameters of probability distributions and UMFs) are relaxed by considering α-corresponding to α-cut defined in the interval [0.01, 1].

Step 2: Repeat step2 to step5 of Algorithm-II.

Proceeding in this way, a set of cumulative p-boxes is obtained.

Step 3: Take LMFs of all the input IVFNs. Here, also relax the convexity condition of the fuzzy focal elements by considering α-level corresponding to α-cut defined in the interval [0.01, 1].

Step 6: Repeat step2 to step5 of Algorithm-II.

Again proceeding in this way, another set of p-boxes are achieved.

From these two families of p-boxes, IVFNs will be produced. First family generates the UMF of the resultant IVFNs and later set generates the LMF of the resulting IVFNs for different fractiles.

Minimum of 11 $\alpha -$ cuts are enough to obtain the shape of the resultant fuzzy numbers at different fractiles, however to obtain a smooth shape the number of alpha-cuts can be increased.

A THEORETICAL CASE STUDY

Here, an effort has been made to carry out non-cancer human health risk assessment with hypothetical data via the proposed algorithms.

It is well known that people or individual has generally been exposed to pollutant/ or harmful compounds via natural and artificial sources as well. One of the artificial sources is produced water which is one of the most noteworthy sources of waste produced in the manufacturing stage of oil and gas procedures. Once released into the water or sea, lots of heavy metals and other harmful compound are obtained in produced water which may initiate toxicity and bioaccumulation in aquatic living beings (Mofarrah and Hussain, 2010). Although numerous organic and inorganic contaminants available in produced water, in this study only the heavy metal *As* is taken into consideration because of its toxicity and high concentration. Apart from produced *As* occurs in seawater, warm springs, groundwater, rivers, and lakes. The unauthorized use of *As* in different areas leads to the worldwide existence of soluble *As* above acceptable levels of 0.010 mg/L. Continuous intake of such exposure like As by aquatic organism leads to their health effect. Some studies find out that high concentration of As effects liver and kidney more than other organs which disrupts the usual metabolism. Therefore, it is palpable that exposure of As effects the usual biological function which again leads to initiation of diseases (Kumari et al, 2017). Thus, intake of such contaminated fish may cause serious health effects to human being. Hence, release of harmful compounds into the ecology evaluation of health risk is essential to gauge possible impact these harmful compounds may have on human health and biota. Therefore, evaluation of a non cancer health risk is carried out here..

Risk Assessment Model

The well validated general risk assessment model (EPA, 2004) is as follows:

$$CDI = \frac{C_f \times FIR \times FR \times EF \times ED \times CF}{BW \times AT}$$

where

$$C_f = PEC \times BCF$$

The required non-cancer risk model for ingestion of fish is:

$$Risk_{non-cancer} = \frac{CDI}{Rfd}$$

Details of terms of the model are depicted in the Table 1.

In this case study, two scenarios are taken to perform the non cancer health risk assessment using the proposed algorithms.

Scenario I

In this scenario, representation of the uncertain parameters *PEC* is considered as probability distribution with fuzzy parameters while other uncertain parameters *ED*, *FIR* and *BCF* are considered as normal FNs. On the other hand, other components of the model are considered as constant. Necessary data for the evaluation of risk are given in the Table 2.

Algorithm-I has been employed in this scenario, to evaluate the non cancer health risk. Since representations of some uncertain parameters are probability distribution with fuzzy parameters and some other model parameters are normal FNs. In this process, a family of p-boxes are obtained. For instance, if α takes equally spaced

Table 1. Abbreviation of terms

Sl. No.	Abbreviation	Full Form
1.	CDI	Chronic daily intake (mg/kg-day)
2.	FIR	Fish ingestion rate (g/day)
3.	FR	Fraction of fish from contaminated source
4.	EF	Exposure frequency (day/year)
5.	ED	Exposure duration (years)
6.	CF	Conversion factor ($=10^{-9}$)
7.	BW	Body weight (kg)
8.	AT	Averaging time (days)
9.	C_f	Chemical concentration of fish tissue (mg/kg)
10	PEC	Predicted environmental concentration (mg/l)
11.	BCF	Chemical bioaccumulation factor in fish (l/kg)
12.	*Rfd*	Reference dose

Table 2. Necessary data for the evaluation of risk

Model Components	Units	Type	Value/Distribution
AT	Days	Constant	25550
BW	Kg	Constant	70
ED	Years	Fuzzy	[20,30,40]
EF	Days/year	Constant	350
FR	-	Constant	0.5
FIR	g/day	Fuzzy	[160, 170, 180]
CF	-	Constant	1*E*-09
PEC for As	μg/l	Probability Distribution	Normal([5,5.5,6],[1,1.35,1.7])
BCF for As	l/kg	Fuzzy	[30, 45, 60]
Oral Rfd for As	mg/(kg.day)	Constant	3.0E-04

11 values (*i.e.*, $\alpha = [0:0.1:1]$), a total of 11 p-boxes (figure -1) will be obtained. 10 distinct p-boxes corresponds to α-level 0 to 0.9 and while there will be one common p-box (like a CDF) corresponds to 1-cut (i.e., $\alpha=1$). From this family of p-boxes, membership functions (fuzzy numbers) of risk at different fractiles can be generated. For example, at 95th fractile the graphical representation of the resultant fuzzy number is depicted in figure 1.

Figure 1. Cumulative distribution function of non cancer health risk for $\alpha = 0:0.1:1$

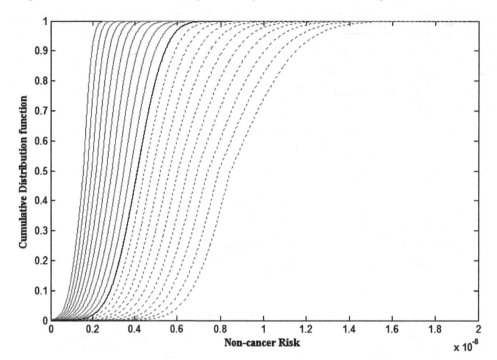

Scenario-II

Here, representation of the uncertain parameters *PEC* and *BCF* are taken as probability distributions with imprecise (BFNs/FNs) parameters. Representations of the other uncertain parameters *ED*, *EF* and *FIR* are Gaussian, Cauchy and triangular fuzzy numbers respectively. Other components of the risk model are taken as constant. Necessary data for the evaluation of risk are depicted in the Table 3.

Algorithm-II has been employed in this scenario, to assess the non cancer health risk. As representations of some uncertain parameters are probability distribution and some other model parameters are Bell-Shaped/normal fuzzy numbers. Here also, a family of p-boxes (table-3) are obtained. For instance, if α takes equally spaced 11 values (*i.e.*, $\alpha = [0.01:0.099:1]$), a total of 11 p-boxes will be obtained. 10 distinct p-boxes corresponds to α-level 0.01 to 0.901 and while there will be one common p-box (like a CDF) corresponds to 1-cut (i.e., α=1). From this family of p-boxes, membership functions (fuzzy numbers) of risk at different fractiles can be generated. Its graphical representation is depicted in figure 4.

Figure 2. Membership function of Risk at 95ᵗʰ fractile

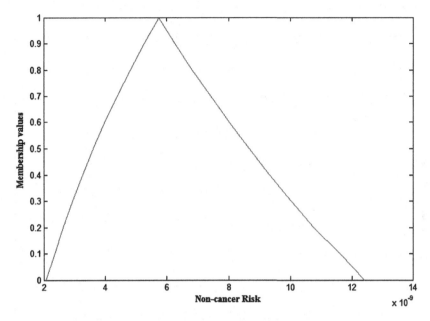

Table 3. Required data for the evaluation of risk

Model Components	Units	Type of Variable	Value/Distribution
AT	Days	Constant	25550
BW	Kg	Constant	70
ED	Years	Fuzzy	Gauss([30,3.3])
EF	Days/year	Fuzzy	Cauchy([350,1])
FR	-	Constant	0.5
FIR	g/day	Fuzzy	[160,170,180]
CF	-	Constant	1E-09
PEC for As	µg/l	Probability distribution	Normal([5,5.5,6],[1.1.35,1.7])
BCF for As	l/kg	Probability distribution	Normal(Gauss(45,1.65), Cauchy(3.5,0.05))
Oral Rfd for As	mg/(kg.day)	Constant	3.0E-04

Scenario-III

Here, representation of the uncertain parameters *PEC* and *BCF* are taken as probability distributions with imprecise (BFNs/FNs) parameters. Representations of the other uncertain parameters *ED*, *EF* and *FIR* are interval valued Gaussian, Cauchy and

Figure 3. Cumulative distribution function of non cancer health risk for $\alpha = [0.01 : 0.099 : 1]$

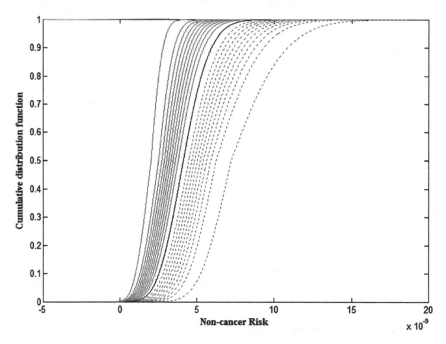

Figure 4. Membership function of Risk at 95th fractile

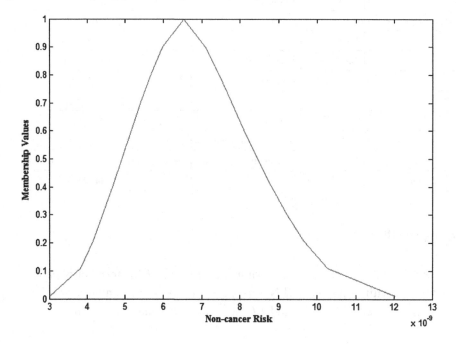

triangular fuzzy numbers respectively. Other parameters remain kept constant. Required data for the calculation of non-cancer human health risk assessment are given in the Table 4.

Since representations some uncertain variables are probability distribution with (Bell-Shaped/normal) fuzzy parameters and on the other hand, representation of some other uncertain models parameters are interval valued Bell-Shaped/normal fuzzy numbers, so algorithm I & II are not appropriate to perform the non cancer risk assessment. Therefore, algorithm III is required to assess non cancer risk here. Using this algorithm-III, two families of p-boxes (figure-5) will be obtained from which IVFNs will be produced (figure-6). First family will generate the UMF of the resultant IVFNs and later will generate the LMF of the resulting IVFNs for different fractiles. Only 11 $\alpha -$ cuts are sufficient to get the rough shape of the resultant fuzzy numbers at different fractiles, however to obtain a smooth shape one may increase number of alpha-cuts.

In a similar fashion like algorithm I & II, from these two families of p-boxes, membership functions of risk at different fractiles can be generated which will be almost bell-shaped interval valued fuzzy numbers. Risk at 95th fractile is calculated and membership functions of risk are interval valued fuzzy numbers depicted in the Figure 6.

Table 4. Required data for the risk assessment

Model Components	Units	Type of Variable	Value/Distribution
AT	Days	Constant	25550
BW	Kg	Constant	70
ED	Years	Fuzzy	Gauss([30,3.3]) UMF Gauss([30,1.65]) LMF
EF	Days/year	Fuzzy	Cauchy([350,1]) UMF Cauchy([350,0.5]) LMF
FR	-	Constant	0.5
FIR	g/day	Fuzzy	[160,170,180] UMF [165,170,175] LMF
CF	-	Constant	1E-09
PEC for As	µg/l	Probability distribution	Normal([5,5.5,6],[1.1.35,1.7])
BCF for As	l/kg	Probability distribution	Normal(Gauss(45,1.65), Cauchy(3.5,0.05))
Oral Rfd for As	mg/(kg.day)	Constant	3.0E-04

Figure 5. Cumulative Belief and plausibility of Risk for $\alpha = 0.01 : 0.099 : 1$

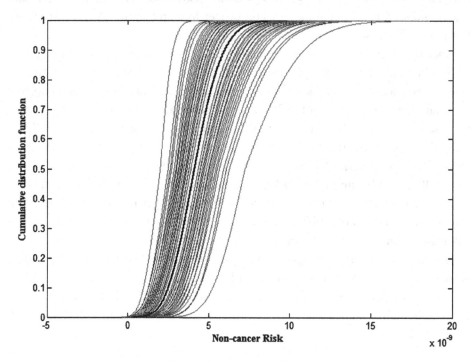

Figure 6. Membership function of Risk at 95th fractile

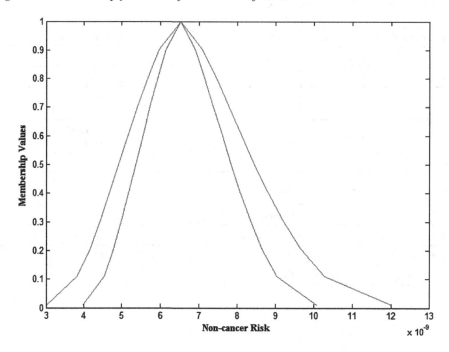

RESULTS AND DISCUSSIONS

Generally, the majority of health risk assessment problem engages treating with uncertainties. More attentiveness requires for all kind of uncertainties and try to include all the necessary information into the assessment process. In the scenario-I, the predicted environmental concentration (*PEC*) is considered as probability distributions with imprecise parameters (i.e., triangular fuzzy mean & standard deviation) while the chemical bioaccumulation factor (*CBF*) considered exposure duration (*ED*) & fish ingestion rate (*FIR*) are considered as triangular fuzzy numbers. In scenario-II, *PEC* and *BCF* are taken as probability distributions with imprecise (BFNs/FNs) parameters. Representations of the other uncertain parameters *ED*, *EF* and *FIR* are Gaussian, Cauchy and triangular fuzzy numbers respectively because of their ability to construe the uncertainty. In scenario-III, *PEC* and *BCF* are taken as probability distributions with imprecise (BFNs/FNs) parameters. Representations of the other uncertain parameters *ED*, *EF* and *FIR* are interval valued Gaussian, Cauchy and triangular fuzzy numbers respectively. Then algorithm-I, II and III have been executed in the entire scenario respectively. The results obtained at different fractiles from 95th to 05th by using algorithm-I, II and III are clearly depicted in Table 5, 6 and 7 respectively. In all the approaches, initial results are obtained as family of p-boxes and finally membership functions of risk are evaluated for different fractiles.

For example, in scenario-I, at 95th fractile, risk is obtained as the fuzzy number around 5.7597e-009 and range of risk is [2.0762e-009, 1.2416e-008]. Similarly, in scenario-II, at 95th fractile, core/modal value of resultant fuzzy number is around 6.5229e-009 and range of the risk value is [3.0339e-009, 1.2012e-008]. Finally, in

Table 5. Fractiles of risk value of the resultant fuzzy numbers (scenario-I)

Fractiles	Core Value	Range
95th	5.7597e-009	[2.0762e-009, 1.2416e-008]
85th	5.1526e-009	[1.8865e-009, 1.0926e-008]
75th	4.7850e-009	[1.7749e-009, 1.0072e-008]
65th	4.4961e-009	[1.6870e-009, 9.3725e-009]
55th	4.2348e-009	[1.6063e-009, 8.7525e-009]
45th	3.9770e-009	[1.4955e-009, 8.2688e-009]
35th	3.7179e-009	[1.3589e-009, 7.9088e-009]
25th	3.4258e-009	[1.2026e-009, 7.4988e-009]
15th	3.0676e-009	[1.0098e-009, 6.9958e-009]
5th	2.4428e-009	[6.8045e-010, 6.1455e-009]

Table 6. Fractiles of risk value of the resultant fuzzy numbers (scenario-II)

Fractiles	Core Value	Range
95th	6.5229e-009	[3.0339e-009, 1.2012e-008]
85th	5.5823e-009	[2.6426e-009, 1.0141e-008]
75th	5.0514e-009	[2.4211e-009, 9.0878e-009]
65th	4.6422e-009	[2.2499e-009, 8.2774e-009]
55th	4.2861e-009	[2.1005e-009, 7.5732e-009]
45th	3.9516e-009	[1.9190e-009, 7.0341e-009]
35th	3.6164e-009	[1.6968e-009, 6.6162e-009]
25th	3.2557e-009	[1.4603e-009, 6.1624e-009]
15th	2.8228e-009	[1.1805e-009, 5.6117e-009]
5th	2.1419e-009	[7.5102e-010, 4.7293e-009]

Table 7. Fractiles of risk value of the resultant fuzzy numbers (scenario-3)

Fractiles	Core Value	Range
95th	6.5229e-009	[3.0339e-009, 1.2012e-008] UMF [3.9686e-009, 1.0079e-008] LMF
85th	5.5823e-009	[2.6426e-009, 1.0141e-008] UMF [3.4568e-009, 8.5088e-009] LMF
75th	5.0514e-009	[2.4211e-009, 9.0878e-009] UMF [3.1670e-009, 7.6249e-009] LMF
65th	4.6422e-009	[2.2499e-009, 8.2774e-009] UMF [2.9430e-009, 6.9450e-009] LMF
55th	4.2861e-009	[2.1005e-009, 7.5732e-009] UMF [2.7477e-009, 6.3541e-009] LMF
45th	3.9516e-009	[1.9190e-009, 7.0341e-009] UMF [2.5103e-009, 5.9018e-009] LMF
35th	3.6164e-009	[1.6968e-009, 6.6162e-009] UMF [2.2196e-009, 5.5512e-009] LMF
25th	3.2557e-009	[1.4603e-009, 6.1624e-009] UMF [1.9102e-009, 5.1704e-009] LMF
15th	2.8228e-009	[1.1805e-009, 5.6117e-009] UMF [1.5442e-009, 4.7083e-009] LMF
5th	2.1419e-009	[7.5102e-010, 4.7293e-009] UMF [9.8241e-010, 3.9680e-009] LMF

scenario-III, at 95th fractile, risk is obtained as the interval valued fuzzy number around 6.5229e-009 and range of risk value of the UMF and LMF of the resulting IVFN are [3.0339e-009, 1.2012e-008] and [3.9686e-009, 1.0079e-008] respectively. As the notion of combination of probability distributions with triangular/bell shaped fuzzy parameters plus different shapes & types of fuzzy numbers to deal with uncertainty is being first observed in human health risk assessment, so no comparative study can be made.

CONCLUSION

In human health risk assessment uncertainties play a crucial rule and dealing with uncertainties is a complex phenomenon. In this article uncertainties are modeled that involved in the risk assessment process/model. Uncertainties are modeled are modeled in the form of probability distributions with imprecise parameters and fuzzy numbers of different types and shapes as well. As different facets of uncertainties rare co-exist in the risk assessment so to carry out risk assessment under uncertain environment, an amalgamate technique through three algorithms are designed. It is found that the proposed technique is novel, efficient and technically sounds. It is further experienced that the technique initially produces a set of P-boxes and from which output risk can be obtained in the form of fuzzy number of different types and shapes.

As a future extension of the work, type-II fuzzy numbers may be considered to address uncertainties and subsequently attempt can be made to perform health risk.

REFERENCES

Abdo, H., & Flaus, J. M. (2016). Uncertainty quantification in dynamic system risk assessment: A new approach with randomness and fuzzy theory. *International Journal of Production Research, 54*(19), 1–24. doi:10.1080/00207543.2016.1184348

Alvarez, D. A. (2006). On the calculation of the bounds of probability of events using infinite random sets. *International Journal of Approximate Reasoning, 43*(3), 241–267. doi:10.1016/j.ijar.2006.04.005

Anoop, M. B., Balaji, R. K., & Lakshmanan, N. (2008). Safety assessment of austenitic steel nuclear power plant pipelines against stress corrosion cracking in the presence of hybrid uncertainties. *International Journal of Pressure Vessels and Piping, 85*(4), 238–247. doi:10.1016/j.ijpvp.2007.09.001

Arunraj, N. S., Mandal, S., & Maiti, J. (2013). Modeling uncertainty in risk assessment: An integrated approach with fuzzy set theory and monte carlo simulation. *Accident; Analysis and Prevention, 55*, 242–255. doi:10.1016/j.aap.2013.03.007 PMID:23567215

Ashraf, M. A., Suzan, A., & Donald, C. W. (2007). Gaussian Versus Cauchy Membership Functions in Fuzzy PSO. *Proceedings of International Joint Conference on Neural Networks.*

Baraldi, P., & Zio, E. (2008). A Combined Monte Carlo and Possibilistic Approach to Uncertainty Propagation in Event Tree Analysis. *Risk Analysis, 28*(5), 1309–1326. doi:10.1111/j.1539-6924.2008.01085.x PMID:18631304

Baudrit, C., & Dubois, D. (2006). Practical Representations of Incomplete Probabilistic Knowledge. *Computational Statistics & Data Analysis, 51*(1), 86–108. doi:10.1016/j.csda.2006.02.009

Baudrit, C., Dubois, D., & Guyonnet, D. (2006). Joint Propagation and Exploitation of Probabilistic and Possibilistic Information in Risk Assessment. *IEEE Transactions on Fuzzy Systems, 14*(5), 593–608. doi:10.1109/TFUZZ.2006.876720

Baudrit, C., Dubois, D., & Perrot, N. (2008). Representing parametric probabilistic models tainted with imprecision. *Fuzzy Sets and Systems, 159*(15), 1913–1928. doi:10.1016/j.fss.2008.02.013

Berleant, D. (1993). Automatically verified reasoning with both intervals and probability density functions. *Interval Computations, 2*, 48–70.

Berleant, D., & Goodman-Strauss, C. (1998). Bounding the Results of Arithmetic Operations on Random Variables of Unknown Dependency Using Intervals. *Reliable Computing, 4*(2), 147–165. doi:10.1023/A:1009933109326

Bojadziev, G., & Bojadziev, M. (1995). *Fuzzy set, Fuzzy logic, application.* Singapore: World Scientific.

Bouissou, O., Goubault, E., Larrecq, J. G., & Putot, S. (2011). A generalization of P-boxes to affine arithmetic. *Computing.* doi:10.100700607-011-0182-8

Chen, Z., Zhao, L., & Lee, K. (2010). Environmental risk assessment of offshore produced water discharges using a hybrid fuzzy-stochastic modeling approach. *Environmental Modelling & Software, 25*(6), 782–792. doi:10.1016/j.envsoft.2010.01.001

Chistropher, H. F., & Bharvikar, R. (2002). Quantification of variability and Uncertainty: A case study of power Plant Hazardous Air Pollutant Emission. In D. Paustenbach (Ed.), *Human Ecological Risk Analysis* (pp. 587–617). New York: John Wiley and Sons.

Conte, S. D., & De Boor, C. (1981). *Elementary Numerical Analysis: An Algorithmic Approach* (3rd ed.). McGraw-Hill.

Crespo, L. G., Kenny, S. P., & Giesy, D. P. (2012). Reliability analysis of polynomial systems subject to P-box uncertainties. *Mechanical Systems and Signal Processing*. doi:10.1016/j.ymssp.2012.08.012

Destercke, S., & Dubois, D. (2009). The role of generalised P-boxes in imprecise probability models. *6th International Symposium on Imprecise Probability: Theories and Applications*, Durham, UK.

Dutta, P. (2013). An approach to deal with aleatory and epistemic uncertainty within the same framework: Case study in risk assessment. *International Journal of Computers and Applications*, *80*(12).

Dutta, P. (2015). Uncertainty modeling in risk assessment based on Dempster–Shafer theory of evidence with generalized fuzzy focal elements. *Fuzzy Information and Engineering*, *7*(1), 15–30. doi:10.1016/j.fiae.2015.03.002

Dutta, P. (2017). Modeling of variability and uncertainty in human health risk assessment. *MethodsX*, *4*, 76–85. doi:10.1016/j.mex.2017.01.005 PMID:28239562

Dutta, P. (2018). Human Health Risk Assessment Under Uncertain Environment and Its SWOT Analysis. *The Open Public Health Journal*, *11*(1), 72–92. doi:10.2174/1874944501811010072

Dutta, P., & Ali, T. (2011). A Proposal for Construction of P-box. *International Journal of Advanced Research in Computer Science*, *2*(5), 221–225.

Dutta, P., & Ali, T. (2015). Aleatory and Epistemic Uncertainty Quantification. In *Applied Mathematics* (pp. 209–217). New Delhi: Springer.

Dutta, P., & Hazarika, G. C. (2017). Construction of families of probability boxes and corresponding membership functions at different fractiles. *Expert Systems: International Journal of Knowledge Engineering and Neural Networks*, *34*(3), e12202. doi:10.1111/exsy.12202

EPA US. (2004). *Risk Assessment Guidance for Superfund, Volume I: Human Health Evaluation Manual (Part E, Supplemental Guidance for Dermal Risk Assessment).* Office of Emergency and Remedial Response, EPA/540/R/99/005, Interim, Review Draft. United States Environmental Protection Agency.

Ferson, S., & Donald, S. (1998). In A. Mosleh & R. A. Bari (Eds.), *Probability Bounds Analysis, Probabilistic Safety Assessment and Management* (pp. 1203–1208). New York, NY: Springer-Verlag.

Ferson, S., Kreinovich, V., Ginzburg, L., Myers, D. S., & Sentz, K. (2003). *Construction of probability boxes and Dempster-Shafer structures.* Tech. Rep. SAND2002-4015, Sandia National Laboratories.

Ferson, S., Nelsen, R., Hajagos, J., Berleant, D., Zhang, J., Tucker, W., ... Oberkampf, W. (2004). *Dependence in probabilistic modelling, Dempster–Shafer theory and probability bounds analysis. Tech. rep.* Sandia National Laboratories.

Flage, R., Baraldi, P., Zio, E., & Aven, T. (2010) Possibility-probability transformation in comparing different approaches to the treatment of epistemic uncertainties in a fault tree analysis. *Reliability, Risk and Safety - Proceedings of the ESREL, Conference,* 714-721.

Flage, R., Baraldi, P., Zio, E., & Aven, T. (2013). Probabilistic and Possibilistic treatment of epistemic uncertainties in fault tree analysis. *Risk Analysis, 33,* 121–133. doi:10.1111/j.1539-6924.2012.01873.x PMID:22831561

Frechet, M. (1935). Generalisations du theoreme des probabilite's totals. *Fundamenta Mathematicae, 25,* 379–387. doi:10.4064/fm-25-1-379-387

Granger, M. M., & Henrion, M. (1992). *Uncertainty- A guide to dealing uncertainty in Quantitative risk and policy analysis.* Cambridge University Press.

Guyonnet, D., Bourgine, B., Dubois, D., Fargier, H., Côme, B., & Chilès, J. P. (2003). Hybrid approach for addressing uncertainty in risk assessments. *Journal of Environmental Engineering, 126,* 68–78. doi:10.1061/(ASCE)0733-9372(2003)129:1(68)

Guyonnet, D., Côme, B., Perrochet, P., & Parriaux, A. (1999). Comparing two methods for addressing uncertainty in risk assessments. *Journal of Environmental Engineering, 125*(7), 660–666. doi:10.1061/(ASCE)0733-9372(1999)125:7(660)

Innal, F., Chebila, M., & Dutuit, Y. (2016). Uncertainty handling in safety instrumented systems according to IEC 61508 and new proposal based on coupling Monte Carlo analysis and fuzzy sets. *Journal of Loss Prevention in the Process Industries, 44,* 503–514. doi:10.1016/j.jlp.2016.07.028

Karami, A., Keiter, S., Hollert, H., & Courtenay, S. C. (2013). Fuzzy logic and adaptive neuro-fuzzy inference system for characterization of contaminant exposure through selected biomarkers in African catfish. *Environmental Science and Pollution Research International, 20*(3), 1586–1595. doi:10.100711356-012-1027-5 PMID:22752811

Karanki, D. R., Kushwaha, H. S., Verma, A. K., & Ajit, S. (2009). Uncertainty analysis based on probability bounds (P-box) approach in probabilistic safety assessment. *Risk Analysis, 29*(5), 662–675. doi:10.1111/j.1539-6924.2009.01221.x PMID:19302279

Kentel, E., & Aral, M. M. (2004). Probabilistic-Fuzzy Health Risk Modeling. *Stochastic Environmental Research and Risk Assessment, 18*(5), 324–338. doi:10.100700477-004-0187-3

Kumari, B., Kumar, V., Sinha, A. K., Ahsan, J., Ghosh, A. K., Wang, H., & DeBoeck, G. (2017). Toxicology of arsenic in fish and aquatic systems. *Environmental Chemistry Letters, 15*(1), 43–64. doi:10.100710311-016-0588-9

Limbourg, P., & de Rocquigny, E. (2010). Uncertainty analysis using evidence theory – confronting level-1 and level-2 approaches with data availability and computational constraints. *Reliability Engineering & System Safety, 95*(5), 550–564. doi:10.1016/j.ress.2010.01.005

Liu, J., Li, Y., Huang, G., Fu, H., Zhang, J., & Cheng, G. (2017). Identification of water quality management policy of watershed system with multiple uncertain interactions using a multi-level-factorial risk-inference-based possibilistic-probabilistic programming approach. *Environmental Science and Pollution Research International,* 1–21. doi:10.100711356-017-9106-2 PMID:28488149

Maxwell, R. M., Pelmulder, S. D., Tompson, A. F. B., & Kastenberg, W. E. (1998). On the development of a new methodology for groundwater-driven health risk assessment. *Water Resources Research, 34*(4), 833–847. doi:10.1029/97WR03605

Modares, M., Mullen, R., L., Muhanna, R., L. (2006). Natural Frequencies of a Structure with Bounded Uncertainty. *Journal of Engineering Mechanics,* 1363-1371.

Mofarrah, A., & Hussain, T. (2010). *Modeling for uncertainty assessment in human health risk quantification: A fuzzy based approach*. International congress on environmental modeling and soft modeling for environment's sake, fifth biennial meeting, Ottawa, Canada.

Nadiri, A. A., Gharekhani, M., Khatibi, R., & Moghaddam, A. A. (2017). Assessment of groundwater vulnerability using supervised committee to combine fuzzy logic models. *Environmental Science and Pollution Research International*, *24*(9), 8562–8577. doi:10.100711356-017-8489-4 PMID:28194673

Pacheco, M. A. C., & Vellasco, M. B. R. (2009). *Intelligent Systems in Oil Field Development under Uncertainty*. Berlin: Springer-Verlag. doi:10.1007/978-3-540-93000-6

Pastoor, T. P., Bachman, A. N., Bell, D. R., Cohen, S. M., Dellarco, M., Dewhurst, I. C., ... Boobis, A. R. (2014). A 21st century roadmap for human health risk assessment. *Critical Reviews in Toxicology*, *44*(S3), 1–5. doi:10.3109/10408444.2014.931923 PMID:25070413

Pedronia, N., Zioa, E., Ferrariob, E., Pasanisid, A., & Couplet, M. (2012). Propagation of aleatory and epistemic uncertainties in the model for the design of a flood protection dike. PSAM 11 & ESREL, Helsinki: Finland.

Pedronia, N., Zioa, E., Ferrariob, E., Pasanisid, A., & Couplet, M. (2013). Hierarchical propagation of probabilistic and non-probabilistic uncertainty in the parameters of a risk model. *Computers & Structures*, *126*, 199–213. doi:10.1016/j.compstruc.2013.02.003

Rębiasz, B., Bartłomiej, G., & Iwona, S. (2017). Joint Treatment of Imprecision and Randomness in the Appraisal of the Effectiveness and Risk of Investment Projects. *Information Systems Architecture and Technology: Proceedings of 37th International Conference on Information Systems Architecture and Technology–ISAT 2016–Part IV*.

Saad, A., Fruehwirth, T., & Gervet, C. (2014). *The P-box CDF-Intervals: Reliable Constraint Reasoning with Quantifiable Information, to appear in Theory and Practice of Logic Programming*. TPLP.

Sambuc, R. (1975). *Function Φ-Flous, Application a l'aide au Diagnostic en Pathologie Thyroidienne, These de Doctoraten Medicine*. University of Marseille.

Santillana Farakos, S. M., Frank, J. F., & Schaffner, D. W. (2013). Modeling the influence of temperature, water activity and water mobility on the persistence of Salmonella in low-moisture foods. *International Journal of Food Microbiology*, *166*(2), 280–293. doi:10.1016/j.ijfoodmicro.2013.07.007 PMID:23973840

Santillana Farakos, S. M., Pouillot, R., Anderson, N., Johnson, R., Son, I., & Van Doren, J. (2016). Modeling the survival kinetics of Salmonella in tree nuts for use in risk assessment. *International Journal of Food Microbiology, 227*, 41–50. doi:10.1016/j.ijfoodmicro.2016.03.014 PMID:27062527

Santillana Farakos, S. M., Schaffner, D. W., & Frank, J. F. (2014). Predicting survival of Salmonella in low-water activity foods: An analysis of literature data. *Journal of Food Protection, 77*(9), 1448–1461. doi:10.4315/0362-028X.JFP-14-013 PMID:25198835

Tucker, W. T., & Ferson, S. (2003). *Probability bounds analysis in environmental risk assessment*. Applied Biomathematics.

Vose, D. (2000). *Risk Analysis-A Quantitative Guide* (2nd ed.). Chichester, UK: John Wiley and Sons Ltd.

Zadeh, L. A. (1965). Fuzzy Set. *Information and Control, 8*(3), 338–353. doi:10.1016/S0019-9958(65)90241-X

Zhang, X., Sun, M., Wang, N., Huo, Z., & Huang, G. (2016). Risk assessment of shallow groundwater contamination under irrigation and fertilization conditions. *Environmental Earth Sciences, 75*, 1–11.

Zwietering, M. H. (2015). Risk assessment and risk management for safe foods: Assessment needs inclusion of variability and uncertainty, management needs discrete decisions. *International Journal of Food Microbiology, 213*, 118–123. doi:10.1016/j.ijfoodmicro.2015.03.032 PMID:25890788

Chapter 6
Green Computing:
A Step Towards Eco-Friendly Computing

Sunil Kumar Mohapatra
College of Engineering and Technology, India

Priyadarshini Nayak
Central Institute of Plastic Engineering and Technology, India

Sushruta Mishra
KIIT University, India

Sukant Kishoro Bisoy
C. V. Raman College of Engineering, India

ABSTRACT

With the increase in the number of computers, the amount of energy consumed by them is on a significant rise, which in turn is increasing carbon content in atmosphere. With the realization of this problem, measures are being taken to minimize the power usage of computers. The solution is green computing. It is the efficient utilization of computing resources while minimizing environmental impact and ensuring both economic and social benefits. Green computing is a balanced and sustainable approach towards achieving a healthier and safer environment without compromising the technological needs of the current and future generations. This chapter studies the architectural aspects, the scope, and the applications of green computing. The emphasis of this study is on current trends in green computing, challenges in the field of green computing, and the future trends of green computing.

DOI: 10.4018/978-1-5225-5793-7.ch006

INTRODUCTION

Green computing, or green IT, is the approach to use computers and other electronic subsystems like - monitors, printer, storage devices, networking and communication systems is a sustainable, efficient and effective way that minimizes harmful impact on the environment. It signifies on ways of practicing and implementing advanced methodologies of using computing resources in an environment amiable way while retaining the overall computing performance (Kotwani, n.d.). It is required to find a way to handle computers and its devices for protecting the environment and society from such E-hazards. Green IT aims at maximizing energy efficiency during the lifetime of the product while reducing the use of hazardous materials, encouraging the recyclability and biodegradability of defunct products and waste materials (Shibly, 2015). Some of the key initiatives towards Green computing can be equipment recycling, reduction of paper usage, virtualization, cloud computing, power management, green manufacturing, etc. In the past few years, attention in Green Computing has motivated research into developing energy-conserving techniques from desktops as well as large enterprise client and server systems. It has even triggered changes in government policies to encourage recycling and lowering energy use by individuals and organizations. Various regulations and laws related to environmental norms forces manufacturers of IT equipments to meet various energy requirements (Wikipedia, n.d.). It is not just eco-friendly but is also much cost effective.

METRICS OF GREEN COMPUTING

In order to reduce the amount of energy consumption by data centres we first need metrics to measure the total power consumption while the system is in operational mode. It helps for determining the power consumed by the running data centers and also inform regarding the scope of reducing it (Sharama & Sharma, 2016). Green Grid has put an effort in this direction (Dillon, Wu & Chang, 2010) by proposing few such metrics. They are as follows:

Power Usage (PU)

PU is a metric that determines the ratio of energy consumed and spent on overhead (Uddin & Rahman, 2012). For an ideal system the value of PU is considered to be 1.0, that indicates 100% efficiency (Ketankumar, Verma, et al., 2015), otherwise the entire electrical power is assumed to be used by the equipments only (Chowdhury, Chatterjee, et al., 2013).

Green Power Usage Effectiveness (GPUE)

The metric PU has been known to have a few drawbacks, such as it is dependent on the time and location, ignores certain factors in order to reduce its rating, applicable only on the dedicated data centers and can't be used for the comparison of data centers. However, the renowned organization named Green Cloud has given the GPUE metric that can overcome the drawbacks of the metric PU. The metric GPUE is more practical to use and therefore yields values from a wider range and higher resolution in comparison to PU (Chowdhury, Chatterjee, et al., 2013).

APPROACHES TO GREEN COMPUTING

Product Longevity

One of the major contributions to green computing is the longevity of product life cycle. As per Life Cycle Assessment report on product life cycle it is clear that the product durability or longevity is one of the best approaches for achieving Green Computing objectives. The manufacturing process accounts for about 70% of the resources used during the life cycle of a system (Shibly, 2015).Increasing the product life will achieve more utilization of resources and also control the excessive manufacturing of unnecessary products. Increasing the life span of any equipment includes the process of upgradability and modularity. The government regulations must push the manufacturers to make more efforts to enhance the product life.

Software and Deployment Optimization

Saving energy in terms of algorithmic efficiency, resource allocation, and virtualization and terminal servers by optimizing the software and deploying it can be considered as an efficient way.

Algorithmic Efficiency

In today's world, the use of search engines to fetch answers to our queries is a common part of the daily life of an individual. Most people believe that there are no environmental disorders caused due to a mere search. But according to a study, an average search releases about 0.2 grams to 7 grams of carbon dioxide (CO_2) each time we search something. Similarly, some of the frequently used software packages require about 70 times more memory(RAM) for performing very basic tasks as compared to what it was some 10 years ago (Shibly, 2015).Hence what we require is a Fine-grained green computing architecture that would run algorithms efficiently and effectively by controlling power consumption on each computing resources.

Resource Allocation

Resource allocation is the approach taken to assign the available resources to various terminals in an economic way. This includes the scheduling of activities and the distribution of resources among the activities considering resource availability and time. Resource allocation can have strategic in nature. In strategic resource allocation method planning is undertaken in a long term process in order to use the available resources to achieve goals for the future. The main objective is the efficient distribution of resources among all the terminals so as to facilitate job completion within the least possible time and incurring less energy and cost.

Virtualization

An effective approach to Green Computing would be the reduction in the number of resources such as storage devices, printers, etc. This involves the reduction of the amount of power needed; the heat produced and thereby would lead to the reduction in the operational and administrative costs of tasks like backup, archival storage etc. Virtualization can be defined as the abstraction of computer resources obtained by running two or more logical systems on one set of physical hardware. Through this concept a system administrator would be able to combine multiple systems into virtual machines utilizing one single, powerful system. One of the primary goals of virtualization is utilizing the available resources to the fullest and hence reducing the power and cooling consumption side effects.

Terminal Servers

Terminal servers are another form of virtualization. Terminal servers refer to the use of a terminal connected to a central server; all though the end user experiences the operating system on his/her own terminal, but the actual operation is being done on the server. It can even be combined with thin clients that use around $1/8^{th}$ of the amount of energy of a workstation, thereby decreasing the energy costs and consumption

Power Management

Power management is a desired feature in every system because it not only reduces the operating costs but also provides prolonged battery life for systems, reduces the energy requirement for cooling and reduces the unwanted noise. The lower the power consumption, the lower will be the heat dissipation, therefore increasing the system stability and reducing the cost and impact on the surrounding. As per this open standard after certain duration of inactivity, the system automatically turns off components such as monitors and hard drives. It also lets a system hibernate by turning off components like the CPU and the RAM during longer period of inactivity.

Power Supply

Power supplies in almost all systems utilize more power than required for the normal functioning of its different components. This not only leads to higher bills but also has hazardous environmental impacts. Certain certification systems for power supply manufacturers like the "80 Plus program" provides standards that must be met in order for the power supply to be efficient (Kotwani, n.d.) .According to this certification, any power supply that meets the standard would use only the power it needs at a given load at any point of time. Initiatives are being undertaken to enhance the efficiency of power supplies thereby promoting energy saving and reducing greenhouse gas emissions.

Storage

The amount of power consumed also depends on the type of storage device that our system uses based on the cost, performance, and capacity. Focusing on reducing the power consumption go hand in hand with reducing RAM idle mode for low power through fixed rotation speed. One of the conventional choices is to use the 3.5" desktop hard disk with the highest possible capacity along good performance (Kotwani, n.d.) . Secondly, a 2.5" laptop hard disk comes into picture because of their ability to consume less power as compared to larger disks. Other than these options, one can go for a solid statehard drive (SSD) that not only draws less than one-third the power of a 2.5" disk, but also are faster and energy efficient.

Displays

The conventional CRT displays were one of the most power consuming components of the entire system. LCD monitors uses cold-cathode fluorescent bulb to generate light and they consume three times less energy in active state, and about ten times less energy in inactive mode as compared to CRTs. LCD screens are not only energy saving but are also soothing to eyes. Light Emitting Diodes (LEDs), are also a popular choice of display that helps in reducing the amount of energy used by the screen.

Materials Recycling

Electronics wastes are a major source of toxins and carcinogens such as lead, mercury, chromium, etc. Recycling of obsolete computers and other electronic components can help in preventing harmful materials and hexavalent compositions out of the environment. Additionally used objects like cartridges, printing papers, and batteries may also be recycled. These recyclable products can be great source of raw materials for manufacturing new ones.

Telecommuting

A lot of greenhouse gas emissions are caused due to travelling through various modes of transport. Generally office workers have been observed to be the frequent travelers in order to attend meeting, conferences, etc. This not just pollutes environment due to vehicle emissions but also increases overhead costing for requirement of space, lighting, heat for fooding, etc. Teleconferencing and video conferencing technologies are a much preferred way than traveling great distances, as it can contribute to the initiatives of green computing. All members can communicate with each other through video and audio systems as if they are in the same room. This brings enormous time as well as cost benefits, and reduces the impact of travelling by cutting down the carbon emissions, on the environment.

GREEN COMPUTING ARCHITECTURAL MODELS

A Green Computing Architecture for Environmental Data Acquisition in IT Data Center

The IT/automation industry can significantly impact on the worldwide carbon foot print by adopting a 'greener' approach to computing for environmental sustainability. In this regard, a novel Smart Green architecture for computing process regulation is there for leveraging on the functionality of Wireless Sensor Network (WSN) for Temperature, humidity and energy data acquisition while offering recommendations on potential energy cost savings for the IT/automation industry. Green computing refers to the practice of using computing resources more efficiently while maintaining or increasing overall performance and ensuring environmental sustainability. Sustainable IT/automation services require the integration of green computing practices such as power management, virtualization, improving cooling technology, recycling, electronic waste disposal, optimization of the IT infrastructure to meet sustainability requirements and also includes the dimensions of environmental sustainability, the economics of energy efficiency, and the total cost of ownership, which includes the cost of disposal and recycling (Okafor, Udeze, Ugwoke, Okoro & Akura, 2013).

Sustainable Technology Strategy (Architectural Journal, n.d.)

Whenever new technology in industry helps growing efficiency into the IT infrastructure then sustainable technology strategy is required in IT architecture and process. Environmental quality metrics are needed to be built in every part of the IT architectural process. The sustainable data center design is implemented as:

- Reusability of IT

Figure 1. Cloud and environmental sustainability
(Garg & Buyya, 2012)

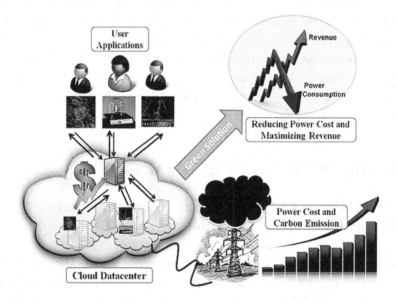

- Complexity Reduction
- Optimization of functional and non-functional Requirements
- Spending of organization's wealth properly

Energy Consumption

Industry should be realized that the source and amount of their energy consumption significantly contributes to greenhouse gas (GHG) emissions. In response to this awareness, organizations are currently using the
Following equation:

Reduced energy consumption = Reduced greenhouse gas emissions

= Reduced operational costs for the data center and business

The technologies that exist for measuring energy consumption and thermal output for data center elements are: Circuit meters, Power strip meters, Base board controller energy consumption metering, external thermal sensor metering, internal server thermal metering

Extended Architecture (Architecture Journal, n.d.)

The above architecture represents environmental impact factors. Due to the proprietary nature of most environmental metering interfaces from separate vendors, IT architects should aggregate these communication models into extensible communication architecture.

1. **Proprietary Energy API Services:** Most vendors have their own proprietary API model to interface with the metering proprietary environmental metering interface systems (as well as versioning changes issues), organizations should assemble a common communication bus devices. Because energy metering architectures differ with many data centers, larger organizations may have more than one proprietary interface environment. It is important to set up are liable design standard that reaches across data centers and technologies.
2. **Environmental Consumption Communication Bus:** A common interface model for different metering and reporting systems for environmental monitoring.
3. **Environmental Consumption Data Aggregation Zone:** This is a common data collection model designed for frequent updatation. This area is the data collection point data centers.

Figure 2. Designing a common environmental metering environment

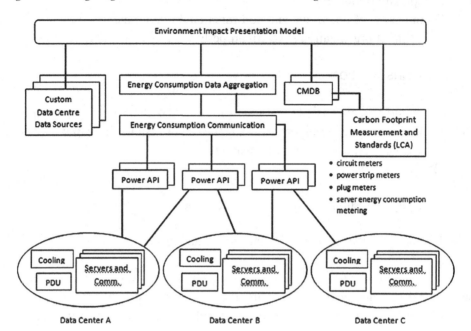

4. **Configuration Management Database Environment (CMDB):** CMDB environments store mostly infrequently updated system configuration information and also able to associate this information with metering systems to understand the impact of configuration decisions on environmental metrics.

5. **Custom Data Center Data Sources***:* In designing a common environmental metering Environment, there often is unique data sources that are important for the solution. Examples include the price and source of energy for that specific data center, operational logistics data, and common data center performance data. It is usually best to keep these systems separate with some common interface standards rather than grouping them together.

6. **Environmental Impact Presentation Model:** This is the presentation aggregation point for different user. While the architectural principles are the same, the architect can leverage many different design options to accomplish the task. Clouds are essentially virtualized data centers and applications offered as services on a subscription basis as shown in Figure1. As Cloud data centers are located in different geographical regions, they have different CO_2 emission rates and energy costs depending on regional constraints. Each data center is responsible for updating this information to Carbon Emission Directory for facilitating the energy-efficient scheduling. A green initiative in

data center designs and other automation processes will facilitate environmental sustainability (Okafor, Udeze, Ugwoke, Okoro & Akura, 2013).

Today, a typical data center with 1000 racks need 10 Megawatt of power to operate, which results in higher operational cost. Thus, for a data center, the energy cost is a significant component of its operating and up-front costs. In addition, in April 2007, Gartner estimated that the Information and Communication Technologies (ICT) industry generates about 2% of the total global CO_2 emissions, which is equal to the aviation industry (Masood Anwar, Syed Furqan Qadri and Ahsan Raza Sattar, 2013). According to a report published by the European Union, a decrease in emission volume of 15% to 30% is required before year 2020 to keep the global temperature increase below 20^C. Thus, energy consumption and carbon emission by Cloud infrastructures has become a key environmental concern. Even the most efficiently built data center with the highest utilization rates will only mitigate, rather than eliminate, harmful CO_2 emissions. The reason given is that Cloud providers are more interested in electricity cost reduction rather than carbon emission. The data collected by the study is presented in Table 2. Clearly, none of the cloud data center in the table can be called as green. (Garg & Buyya, 2012)

Table 1. Percent of power consumption by each data center device

Data Center Devices	Percent of Power Consumption
Cooling devices (Chillier, Computer Room Air Conditioning (CRAC)	33% +9%
IT Equipment	30%
Electrical Equipment (UPS, Power Distribution Units (PDUs), lighting)	28%

(Garg & Buyya, 2012)

Table 2. Comparison of significant cloud data centers

Cloud Data Centers	Location	Estimated Power Usage Effectiveness	% of Dirty Energy Generation	% of Renewable Electricity
Google	Lenoir	1.21	50.5% Coal, 38.7% Nuclear	3.8%
Apple	Apple, NC	-----	50.5% Coal, 38.7% Nuclear	3.8%
Microsoft	Chicago, IL	1.22	72.8% Coal, 22.3% Nuclear	1.1%
Yahoo	La Vista, NE	1.16	73.1% Coal, 14.6% Nuclear	7%

(Garg & Buyya, 2012)

The Green Cloud framework, which takes into account these goals of provider while curbing the energy consumption of Clouds. The high level view of the green Cloud architecture is given in Figure 3. The goal of this architecture is to make Cloud green from both user and provider's perspective. In the Green Cloud architecture, users submit their Cloud service requests through a new middleware Green Broker that manages the selection of the greenest Cloud provider to serve the user's request. A user service request can be of three types i.e., software, platform or infrastructure. The Carbon Emission Directory maintains all the data related to energy efficiency of Cloud service. Then, it selects the set of services that will result in least carbon emission and buy these services on behalf users. The Green Cloud framework is designed such that it keeps track of overall energy usage of serving a user request. It relies on two main components, Carbon Emission Directory and Green Cloud offers, which keep track of energy efficiency of each Cloud provider and also give incentive to Cloud providers to make their service "Green". In general, a user can use Cloud to access any ofthese three types of services (SaaS, PaaS, and IaaS), and therefore process of serving them should also be energy efficient.

Figure 3. Green cloud architecture
(Garg & Buyya, 2012)

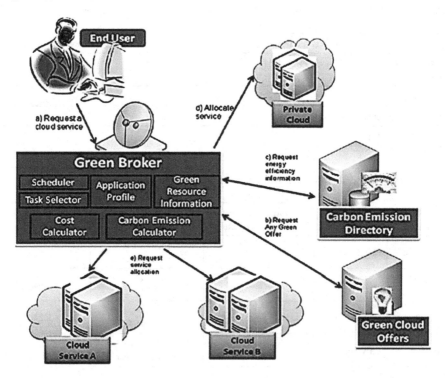

- **SaaS Level:** SaaS providers basically offers software installed on their own data centers or resources from IaaS providers, the major role of SaaS is to design and measure energy efficiency of their software design, implementation, and deployment.

- **PaaS Level:** PaaS providers offer in general the platform services for application development. The platform facilitates the development of applications which ensures system wide energy efficiency. In addition, platforms itself can be designed to have various code level optimizations which can cooperate with underlying complier in energy efficient execution of applications.

- **IaaS Level:** Providers in this layer plays most crucial role in the success of whole Green Architecture since IaaS level not only offer independent infrastructure services but also support other services offered by Clouds. They use latest technologies for IT and cooling systems to have most energy efficient infrastructure. By using virtualization and consolidation, the energy consumption is further reduced by switching-off unutilized server. Various energy meters and sensors are installed to calculate the current energy efficiency of each IaaS providers and their sites. This information is advertised regularly by Cloud providers in Carbon Emission Directory.

In order to validate the framework and to prove that it achieves better efficiency in terms of carbon emission, five policies (Green and profit-oriented) have employed for scheduling by Green Broker (Retrieved from www.architecturejournal.net).

1. **Greedy Minimum Carbon Emission (GMCE):** In this policy, user applications are assigned to Cloud providers in greedy manner based on their carbon emission.

2. **Minimum Carbon Emission - Minimum Carbon Emission (MCE-MCE):** This is a double greedy policy where applications are assigned to the Cloud providers with minimum Carbon. Emission due to their data center location and Carbon emission due to application execution.

3. **Greedy Maximum Profit (GMP):** In this policy, user applications are assigned in greedy manner to a provider who execute the application fastest and get maximum profit.

4. **Maximum Profit - Maximum Profit (MP-MP):** This is double greedy policy considering profit made by Cloud providers and application finishes by its deadline.

5. **Minimizing Carbon Emission and Maximizing Profit (MCE-MP):** In this policy, the broker tries to schedule the applications to those providers which results in minimization of total carbon emission and maximization of profit.

Wireless Sensor Network for Data Center Cooling (Liu, Zhao, O'Reilly, Souarez, Manos, Liang & Terzis, 2010)

Wireless sensor network (WSN) is the best technology for monitoring the task processing as its low-cost and can provide wide range coverage, and can be easily deployed over a geographical area. It does not require any additional network and facility infrastructure in an IT data center environment. A data center can have several adjacent server rooms and each room can have several hundred of racks. We have observed that up to 5°C temperature variation with a couple of distance. Deploying multiple sensing points per rack provides thousands of sensors in the facility. Number of sensors are within the communication range (10 to 50 meters for IEEE 802.15.4 radios, for example), leading to high packet collision probabilities. The overall goal is to understand how energy is consumed in data centers as a function of design, cooling supply and server hardware with workload distribution, data collection and analysis to use this understanding for optimizing and control the data center resources.

System Overview

The figure below is represented the architecture of the Data Center Genome system using both physical and cyber properties of a data center and The tools needed for facility implementation and performance optimization include the following Key components like: Facility layout, Cooling system, Power system, Server performance,

Figure 4. An illustration of the cross section of a data center. Cold air is blown from floor vents, while hot air rises from hot aisles. Mixed air eventually returns to the CRAC where it is cooled with the help of chilled water, and the cycle repeats.

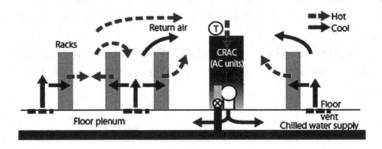

Figure 5. The thermal image of a cold aisle in a data center. The infrared thermal image shows significant variations on intake air temperature across racks and at different heights.

Load variation, Environmental conditions, Real-time monitoring and control, Change management, Capacity planning, Dynamic server provisioning and load distribution, Fault diagnostics and fault tolerance.

Architecture for Green Computing in Virtualization

Virtualization is the abstraction of an OS and applications running on it from the hardware. Physical resources can be split into a number of logical slices called Virtual Machines(VMs).Each VM can accommodate an individual OS creating for the user a view of a dedicated physical resource and ensuring performance and failure isolation between VMs sharing a single physical machine. The virtualization layer lies between the hardware and OS Therefore, a Virtual Machine Monitor (VMM) tacks control over resources and has to be involved in the system's power management in order to provide efficient operation (Sonu Choudhary, 2014).

virtualization reduces carbon emissions due to lesser amount of materials consumed in manufacturing for instance, 'zero clients' contain no processor, memory or other moving elements(hard disks). E-waste reduction, as opposed to e-waste recycling, is a step towards greener environment. 'Zero-clients' can last for 8-10 years are opposed to 3-4 years for a conventional PC. Virtualization helps reduce the ecological impact at all stages of computer lifecycle. IT department also needs to measure the green benefits that accrue from their virtualization efforts in organization. This data is vital to articulate the savings to technical and business decision makers in the organization.

The Different Types of Virtualization for Green Computing are:

1.   Hardware Virtualization
2.   Server Virtualization

Figure 6. Overall architecture for the Data Center Genome system. Data collected from physical and cyber systems in data centers is correlated and analyzed to provide models and Tools for data center management and performance optimization.

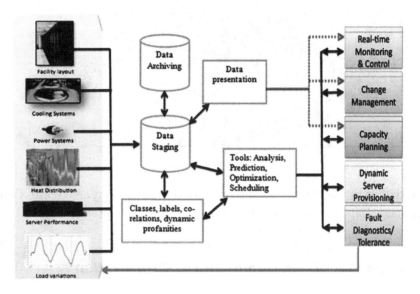

Figure 7. Traditional vs. virtual architecture

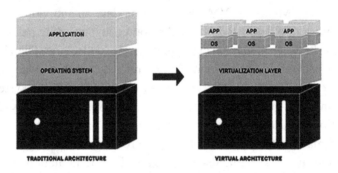

Figure 8. Virtualization

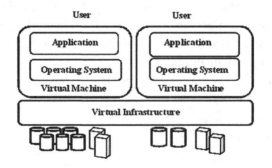

3.   Network Virtualization
4.   Application Virtualization
5.   Hardware Virtualization
6.   Storage Virtualization

Hardware Virtualization

Hardware virtualization is the virtualization of computers on hardware platforms, certain logical abstractions of their componentry, or only the functionality required to run various operating systems. A hypervisor or virtual machine monitor (VMM) is computer software, firmware or hardware that creates and runs virtual machines. A computer on which a hypervisor runs one or more virtual machines is called a *host machine*, and each virtual machine is called a *guest machine*. The hypervisor presents the guest operating systems with a virtual operating platform and manages the execution of the guest operating systems. Multiple instances of a variety of operating systems may share the virtualized hardware resources: for example, Linux, Windows, and mac OS instances can all run on a single physical x86 machine. (Wikipedia – Hardware Virtualization, n.d.)

Server Virtualization (Tiwari, 2015)

Server virtualization offers a way to consolidate servers by allowing us to run multiple different workloads on one physical host server. A "virtual server" is a software implementation that executes programs like a real server. Multiple virtual servers

Figure 9. Hardware virtualization

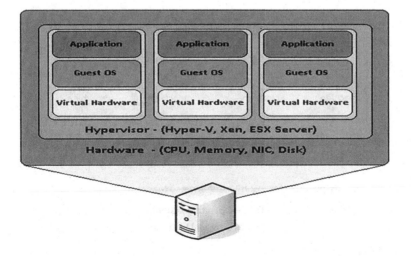

can work simultaneously on one physical host server. Therefore, instead of operating many servers at low utilization, virtualization combines the processing power onto fewer servers that operate at higher total utilization. Some virtualization solutions also have built-in resiliency features, such as high availability, load balancing and failover capabilities. Due to these benefits, virtualization has become commonplace in large data centers. A 2011 survey of over 500 large enterprise data centers found that 92% use virtualization to some degree. (Malviya et al. 2013) Of those, the ratio of virtual servers to physical host server averaged 6.3 to 1 and 39% of all servers were virtual. However, virtualization is less common in small data centers. A 2012 NRDC paper entitled *Small Server Rooms; Big Energy Savings* (Kamdar, IIT Kharagpur) included an informal survey of 30 small businesses (ranging from 3 to 750 employees) and found that only 37% used virtualization.

Network Virtualization

In computing, network virtualization or network virtualization is the process of combining hardware and software network resources and network functionality into a single, software-based administrative entity, a virtual network. (Wikipedia – Network Virtualization, n.d.)

In brief, following are some of the benefits of network virtualization:

- It helps in de-ossifying the current network architectures.
- It allows multiple virtual networks to coexist over a shared physical infrastructure.

Figure 10. Server virtualization in data centers

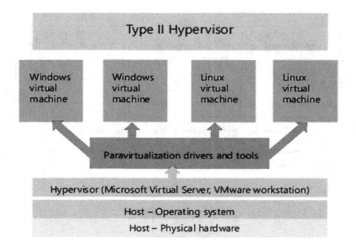

Figure 11. Network virtualization in data centers

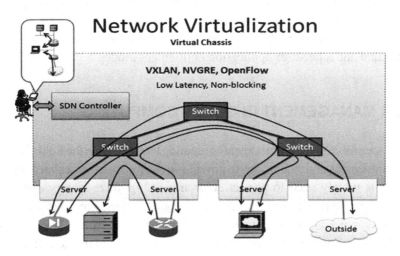

- It provides paths to the future networks approaches.
- It allows the deployment of new business roles and players.
- It reduces/shares cost of ownership.
- It optimizes the resource (network infrastructure) usage.

Application Virtualization

In conversations with customers during workshop sessions we regularly receive the questions: Why do we need application virtualization, what are the benefits and downsides of application virtualization and what is the difference between application deployment and virtualization? The primary reasons for implementing application virtualization are:

Applications are quickly and easy delivered.

- It is simple and easy to upgrade applications.
- There is no need to "install" applications anymore.
- Reduce regression testing time.
- Allow multiple versions of the same application to be run simultaneously on multiple versions of Windows Operating System, greatly reducing the number of server silos.
- Allow non-multiuser versions to run simultaneously in a session virtualization environment.
- Application virtualization is an essential part in 'layering' of OS | Applications | user configuration.

- It's an important component in the complete application and desktop delivery stack.
- Improves end-user mobility – access personalized applications from any machine and a per-user application entitlement model.

POWER MANAGEMENT IN GREEN COMPUTING

Green computing is one of the emergent computing technology in the field of computer science engineering and technology to provide Green Information technology (Green TI/GC). It is mainly used to protect environment, optimize energy consumption and keeps Green environment. Increasing energy efficiency and reducing the use of hazardous materials are the main goals of green computing. Green computing ultimately focuses on ways in reducing overall environmental impacts. It require the integration of Green computing Practices such as recycling, electronic waste removal power consumption, virtualization, improving cooling technology, and optimization of the requirements. The major power consumption components are processors and the main memory in the servers. Green computing is the concept which is trying

Figure 12. Application virtualization

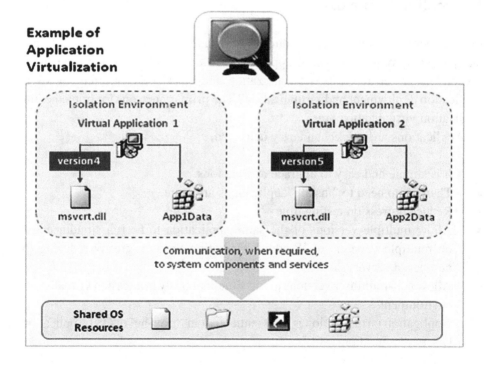

to confine this procedure by inventing new methods that would work efficiently while consuming less energy and making less population. Power management for computer systems are desired for many reasons, particularly:

- Prolong battery life for portable and embedded systems.
- Reduce cooling requirements.
- Reduce noise.
- Reduce operating costs for energy and cooling.

Green Computing Techniques for Power Management in Computer System

These techniques can be classified at different levels(see Figure 13):

1. Hardware and Firmware Level
2. Operating System Level
3. Virtualization Level
4. Data Center Level

Hardware and Firmware Level

The DPM techniques applied at the hardware and firmware Level can be divided in to two categories:

1. Dynamic Components Deactivation (DCD).
2. Dynamic Performance Scaling (DPS).

The DCD techniques are built upon the idea of the clock gating of parts of an electronic components or complete disabling during periods of inactivity. The problem

Figure 13. Power management techniques in green computing

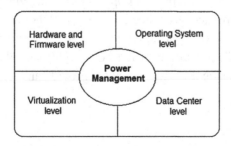

could be easily solved if transitions between power states would cause negligible power and performance overhead. However transitions to low-power states usually lead to additional power consumption and delays caused by the re-initialization of the components. A transition to low-power state is worthwhile only if the period of inactivity is longer than the aggregated delay of transitions form and into the active state and saved power is higher than required to reinitialize the components.

Dynamic Component Deactivation (DCD)

Computer components that do not support performance scaling and can only be deactivated idle. The problem is trivial in the case of a negligible transition has to be done only if the idle period is transitions lead not only to delays which can degrade performance of the systems, but to additional power. Therefore to achieve efficiency a transition has to be done only if the idle period is long enough to cover the transition overhead. DCD techniques can be divided into predictive and stochastic. Predictive techniques are based on the correlation between the past history of the system behavior and its near future. A non-ideal prediction can result in an over-prediction or under-prediction. An over-prediction means that the actual idle period is shorter than the predicted leading to a performance penalty. On the other hand, an under-prediction means that the actual idle period is longer the predicted. Predictive techniques can be further split into static and adaptive, which are discussed below Static techniques utilize some threshold for a real-time execution parameter to make predictions of idle periods. The simplest policy is called fixed timeout. The idea is to define the length of time after which a period of inactivity can be treated as long enough to do a transition to a low-power state. Activation of the component is initiated once the first request to a component is received. Another way to provide the adaptation is to maintain a list of possible values of the parameter of interest and assign weights to the values according to their efficiency at previous intervals. The actual value is obtained as a weighted average over all the values in the list. In general, adaptive techniques are more efficient than static when the type of the workload is unknown. Another way to deal with non-deterministic system behavior is to formulate the problem as a stochastic optimization, which requires building of an appropriate probabilistic model of the system. It is important to note, that the results, obtained using the stochastic approach, are expected values, and there is no guarantee that the solution will be optimal for a particular case. Moreover, constructing a stochastic model of the system in practice may not be straightforward. If the model is not accurate, the policies using this model may not provide an efficient system control.

Dynamic Performance Scaling (DPS)

DPS includes different techniques that can be applied to computer components supporting dynamic adjustment of their performance proportionally to the power consumption. This idea lies in the roots of the widely adopted Dynamic Voltage and Frequency Scaling (DVFS) technique. DVFS reduces the number of instructions a processor can issue in a given amount of time, thus reducing the performance. This, in turn, increases run time for program segments which are sufficiently CPU-bound.

Operating System Level

The characteristics used to classify the operating system levels are shown in Table 3.

Virtualization Level

The virtualization level enables the abstraction of an OS and applications running on it from the hardware. The virtualization layer lies between the hardware and OS and; therefore, a Virtual Machine Monitor (VMM) takes control over resource multiplexing and has to be involved in the system's power management in order to provide efficient operation. There are two ways of how a VMM can participate in the power management:

1. A VMM can act as a power-aware OS without distinction between VMs: monitor the overall system's performance and appropriately apply DVFS or any DCD techniques to the system components.
2. Another way is to leverage OS's specific power management policies and application-level knowledge, and max power management calls from different VMs on actual changes in the hardware's power state or enforce system wide power limits in a coordinated manner.

Table 3. Characteristics of operating system level

Project Name	System Resources	Target Systems	Goal	Power Saving Techniques
Ondem and Government	CPU	Arbitrary	Minimise power consumption	DVFS
Eco System	CPU, Memory, Disk Storage, Network, Interface	Mobile System	Achieve target Battery lifetime	Resource throttling
Nemesis OS, Neugeba And McAuley	CPU, Memory, Disk storage, Network Interface	Mobile Systems	Achieve Target Battery lifetime	Resource Throttling

Data Center Level

The main goal of data center level is:

- Minimize energy consumption, satisfy performance requirements.
- Minimize power consumption, minimize performance loss.

FUTURE WORK

As discussed earlier the of growth in computing needs energy which tends to global warming which is great challenge for IT industry. The future aspects of Green Computing are going to be based on energy efficiency instead of reduction in consumption (Tiwari, 2015). The fundamental focus of Green IT is in the organization's level to self interest in energy cost reduction at Data Centres and the result of which the reduction in carbon generation. The Green IT needs to focus on beyond the energy use in the Data Centre and improving alignment with overall corporate social responsibility efforts. This focus will demand the development of Green Computing strategies.

Figure 14. Data center energy opportunities

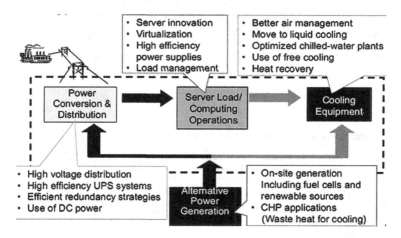

CONCLUSION

Green computing in the future is going to be limitless. As more and more businesses work towards being sustainable the need for processing the data grows expeditiously. So even beyond making computing hardware more green and reducing its carbon footprint, the future will find a larger need for programming to analyze efficiently all of the data needed to track the metrics in order to move forward. Green computing is invaluable in managing the reporting of sustainable project tracking in order to find common problems that are occurring and solutions to these issues. The next wave is Nanodata being created from mobile sensors and controls. Nanodata is about individuals and individual companies to give us insight into how we use equipment. This allows us to figure out ways to do what we want to do, when we want to do it and using the least amount of resources to achieve it. Whether it be software that is strictly analyzing the data looking for new metrics to measure the progress, new software for engineers to use to develop new processes, or software for architects to use when designing new Leed or more sustainable buildings and plants. The capabilities and needs are ever changing and help us not only feel good about our decisions but help to leave a better world for next generation. There are lots of techniques in green computing. By using these techniques we can save energy, emission of CO_2, air pollution and toxic materials. Green computing is not about going out but designing new products in this way to reduce energy consumption. Green computing in future will also help in recycling E-waste and scrap computers.

REFERENCES

Anwar, M., Qadri, S. Q., & Sattar, A. R. (2013). Green Computing and Energy Consumption Issues in the Modern Age. *IOSR Journal of Computer Engineering*, *12*(6), 91–98.

Architecture Journal. (n.d.). Retrieved from: www.architecturejournal.net

Choudhary, S. (2014). A Survey on Green Computing Techniques. *International Journal of Computer Science and Information Technologies*, *5*(5), 6248–6252.

Chowdhury, C.R., Chatterjee, A., Sardar, A., Agarwal, S. & Nath, A. (2013). A Comprehensive study on Cloud Green Computing: To Reduce Carbon Footprints Using Clouds. *International Journal of Advanced Computer Research*, 1-9.

Dillon, T., Wu, C., & Chang, E. (2010). Cloud Computing: Issues and Challenges. *2010 24th IEEE International Conference on Advanced Information Networking and Applications*, 27-33, DOI 10.1109/AINA.2010.187

Garg, S. K., & Buyya, R. (2012). *Green Cloud computing and Environmental Sustainability.* Cloud computing and Distributed Systems (CLOUDS) Laboratory. Retrieved from www.cloudbus.org/papers/Cloud-EnvSustainability2011.pdf

Garg, S. K., & Buyya, R. (n.d.). Green Cloud computing and Environmental Sustainability. Cloud computing and Distributed Systems (CLOUDS). *Laboratory.*

Ketankumar, D. C., Verma, G., & Chandrasekaran, K. (2015). A Green Mechanism Design Approach to Automate Resource Procurement in Cloud. *Procedia Computer Science, 54*, 108 – 117. Doi:10.1016/j.procs.2015.06.013

Kotwai, P. (n.d.). Green computing. *Scribd.* Retrieved from: https://www.scribd.com/document/140778684/51161349-Green-Computing

Liu, J., Zhao, F., O'Reilly, J., Souarez, A., Manos, M., Liang, C.-J. M., & Terzis, A. (2010). *Project Genome: Wireless Sensor Network for Data Center Cooling.* Greenpeace International.

Malviya, P., & Singh, S. (2013). A study about green computing. *International Journal of Advanced Research in Computer Science and Software Engineering, 3*(6), 790-794.

Okafor, K. C., Udeze, C. C., Ugwoke, F. N., Okoro, O. V., & Akura, M. (2013). Smart Green: A Novel Green Computing Architecture for Environmental Data Acquisition in IT/Automation Data Center. *African Journal of Computing and ICT, 6*(5).

Sharma, S., & Sharma, G. (2016). Impact of Energy-Efficient and Eco-Friendly Green Computing. *International Journal of Computer Applications, 143*(7), 20-28. doi:10.5120/ijca2016910250

Shibly, A. (2015). *Green computing: - Emerging issue in IT.* Retrieved from: https://www.researchgate.net/publication/276266707_Green_Computing_-_Emerging_Issue_in_IT

Tiwari, N. (2015). Green Computing. *International Journal of Innovative Computer Science & Engineering, 2*(1).

Uddin, M., & Rahman, A.A. (2012). Energy efficiency and low carbon enabler green IT framework for data centers considering green metrics. *Renewable and Sustainable Energy Reviews, 16*, 4078–4094. doi:10.1016/j.rser.2012.03.014

Wikipedia. (n.d.). Green computing. *Wikipedia.* Retrieved from: http://en.wikipedia.org/wiki/Green_computing

Wikipedia – Hardware Virtualization. (n.d.). Hardware virtualization. *Wikipedia.* Retrieved from: https://en.wikipedia.org/wiki/Hardware_virtualization

Wikipedia – Network Virtualization. (n.d.). *Network Virtualization.* Retrieved form https://en.wikipedia.org/wiki/network_virtualization

Chapter 7

Software Quality Measurement:
State of the Art

Dalila Amara
SMART Lab, Université de Tunis, Institut Supérieur de Gestion, Tunis, Tunisie

Latifa Ben Arfa Rabai
SMART Lab, Université de Tunis, Institut Supérieur de Gestion, Tunis, Tunisie &
College of Business, University of Buraimi, Al Buraimi, Oman

ABSTRACT

Software measurement helps to quantify the quality and the effectiveness of software to find areas of improvement and to provide information needed to make appropriate decisions. In the recent studies, software metrics are widely used for quality assessment. These metrics are divided into two categories: syntactic and semantic. A literature review shows that syntactic ones are widely discussed and are generally used to measure software internal attributes like complexity. It also shows a lack of studies that focus on measuring external attributes like using internal ones. This chapter presents a thorough analysis of most quality measurement concepts. Moreover, it makes a comparative study of object-oriented syntactic metrics to identify their effectiveness for quality assessment and in which phase of the development process these metrics may be used. As reliability is an external attribute, it cannot be measured directly. In this chapter, the authors discuss how reliability can be measured using its correlation with syntactic metrics.

DOI: 10.4018/978-1-5225-5793-7.ch007

INTRODUCTION

Quality is crucial mainly in delicate systems. Poor quality could cause economical and financial loss and it may cause catastrophic situations in organizations using software products. For example, on January 15, 1990, a fault in the software of switching systems has engendered severe perturbations in a long-distance network as well as the shutdown of a number of local phone series (Lyu, 1996). Moreover, on April 19, 2006, the software system of Transportation Security Administration (TSA), U.S. Department of Homeland Security, indicated an image of suspicious device which is a part of routine testing, rather than indicating that a test was underway. Hence, to search the suspicious device, the security area was evacuated for two hours which cause more than 120 flight delays (Wong, 2009). Generally, software quality is represented by different attributes which may be internal i.e size, complexity and cohesion (Helali, 2015) or external i.e reliability, maintainability and portability (Saini et al. 2014).

According to Al-Qutaish (Al-Qutaish. 2010), these attributes are organized in a structured manner by various quality models including McCall's, FURPS, Boehm's and ISO 25010 that replaced the ISO 9126. Among these attributes, reliability was identified as the only common factor of cited models (Farooq et al., 2012).

In the context of quality measurement, numerous models, techniques and metrics have been proposed. For example, different semantic metrics suites are proposed in object oriented (OO) paradigm to measure different dimensions of software quality (Stein et al., 2004; Etzkorn and Delugach, 2000; Mili et al., 2014). Before the emergence of semantic metrics, an important number of syntactic metrics has also been proposed and used for internal software attributes measurement i.e cohesion and complexity (Chidamber and Kemerer 1994; Li 1998). However, little works have investigated the relationship of these metrics with external ones like reliability in order to estimate it. For example Vipin Kumar et al. (Vipin Kumar et al., 2011) proposed a tool called ANTLER based on the C++ programs as input to compute the optimal value of syntactic metrics and to deduce its influence on reliability. They used some OO syntactic metrics to identify which aspects constitute the reliability of the software. Other empirical studies are presented to show the impact of OO properties on quality attributes like maintainability (Abreu and Melo, 1996).

In brief, this chapter aims at presenting the fundamentals of software quality measurement including quality models, techniques and metrics. Moreover, our intention is to present a qualitative approach to measure software reliability as an important software dependability attribute using OO syntactic metrics. This approach will investigate the relationship of these metrics with reliability in order to measure it.

This chapter is structured into five major sections; the authors will present basic elements of software quality including quality attributes and models in section 1.

In section 2 and 3, fundamentals and ways of software quality measurement will be identified. In section 4, a review of the well-known OO syntactic metrics will be presented. Section 5 presents basics of reliability measurement and the impact of internal attributes on it. Finally, the authors conclude and give some perspectives.

QUALITY IN SOFTWARE ENGINEERING

This section presents basics of software quality engineering like quality concept, software quality attributes and models.

Software Quality

Software engineering (SE) is a well-known concept defined by quality standards and famous authors. The most basic common definition identifies SE as a set of engineering activities such as planification, modeling, implementation and testing that support the development of the software (Fenton and Pfleeger 1996).

Software quality is defined through various quality attributes also called features or characteristics. It means the capability of these factors to achieve the implied requirements (IEEE 610.12, 1990; ISO/IEC 25010:2011).

Software Quality Attributes

Numerous definitions of software quality attributes are found in well-known researches and those are most used ones:

- The IEEE standard (IEEE 610.12, 1990) defines an attribute as a characteristic of an item that affects its quality.
- Lyu defines attributes as measures used to assess the performance of the software by describing its behavior (Lyu 1996). Furthermore, it is a property of an entity. For example, the amount of work executed by software is described by its size (Fenton and Bieman 2014).

To sum up, these attributes are required to reflect the software quality and to describe it not only by taking into account the product itself through its structural properties but also its relation with its environment. As consequence, internal and external quality attributes are identified (Fenton and Pfleeger 1996):

Internal Quality Attributes

Internal attributes are structural properties of the code whose quantification depends only on the program representation (Harrison et al. 1998). Bevan defines these attributes as static properties of the code and they include size, complexity, cohesion, coupling, etc (Bevan 1999). Their quantification from the representation of the product makes them early available in the software development life cycle (SDLC) (Fenton and Bieman 2014).

- **Size:** A quality indicator measured by the sum of lines of code that are the executable statements (apart from comments and blanks) (Rosenberg et al. 1998). Simply, it presents the amount of work produced by the program and may be computed without executing it.
- **Complexity:** Appears when a design or implementation of a system or a component is difficult to understand or verify (IEEE 610.12 1990). Complexity is also viewed as problem complexity and algorithm complexity interpretation. In fact, the former is the amount of resources (space or time) needed to obtain an optimal solution to the problem. The latter is the complexity of the algorithm implemented to solve the problem (Fenton and Pfleeger 1996).
- **Cohesion:** The degree to which the tasks performed by a single software module are related to one another (IEEE 610.12 1990). Besides, cohesion describes how the elements of a module are related together and reflects the degree of consistency between them (Fenton and Bieman 2014).
- **Coupling:** Identifies the relationship between the elements of the different system' modules. Whether, there is a stronger coupling between modules, the difficulty to understand, change or correct them goes up which leads to system complexity increase (Briand et al. 1996).

These attributes are important for two reasons; (1) they are available from the first steps of software development, and, (2) they are simple to capture than external attributes. As far as we're concerned to reduce system complexity, classes with high cohesion and low coupling should be created.

External Quality Attributes

External attributes reflect the user' perspective since they are affected by environmental elements to which the product is linked (Harrison et al. 1998; Bevan 1999). By contrast to internal attributes, these attributes are available after the program implementation and execution (Fenton and Bieman 2014). From above definitions, authors note that external attributes require additional information related to the system' environment. For example, Fenton notes that reliability of software as an external attribute depends on both machine environment and the user (Fenton and Bieman 2014).

- **Maintainability:** Means that a software system (or a component) is easy to modify and to adopt to the environment' changes (IEEE 610.12 1990).
- **Flexibility:** The effective and efficient use of a product or a system in different applications and environments (ISO/IEC 25010:2011; IEEE 610.12 1990).
- **Reliability:** The fact to perform the system' required functions under a specified period of time and with respect to the specification (IEEE 610.12 1990).
- **Usability:** The ease of operating, preparing inputs and interpreting the system' outputs by a user (IEEE 610.12, 1990).
- **Portability:** The ability of the developed software to be used different environments (Al-Qutaish 2010).
- **Interoperability:** The exchange and the use of information between different components. Also, it means that the product can be used with other applications (ISO/IEC 2126-1, 2001; Mili and Tchier 2015).
- **Performance:** Both the IEEE standard (IEEE 610.12 1990) and Al-Qutaish (Al-Qutaish 2010) define performance as the ability of a system to accomplish its functions with respect to different constraints like speed and memory storage.
- **Supportability:** The supportability is the capability of the software to still after deployment (Al-Qutaish 2010).

As a conclusion we concentrate on the previous cited attributes since they are the main ones known in literature. We believe that both internal and external attributes are important for quality measurement since they are not disjoint. In this context, Fenton and Bieman emphasize that internal attributes may be used as predictors of the external ones (Fenton and Bieman 2014).

Software Quality Models

Quality models are proposed to describe the interaction between quality attributes to describe them and to facilitate their measurement which allow comparing one product to another. These models decompose quality attributes based on a tree-like structure that consists generally of three major levels. Firstly, the upper branche involves the most important quality factors that we would like to quantify. These factors may be external like reliability and maintainability or internal like testability.

Secondly, each quality factor from the upper branche is structured into other lower-level factors called criteria which are easier to measure than the basic ones. These quality criteria also called sub factors may be internal attributes. Finally, the third level involves directly measurable attributes called metrics and used to measure the lower level criteria (Fenton and Bieman 2014).

Thanks to metrics we can measure the lower-level criteria which in turn allow us to measure the high-level basic factors. For example, for Fenton and Bieman (Fenton and Bieman 2014), the maintainability factor can be described by three sub factors which are correctability, testability and expandability. Different metrics are defined to measure these attributes which are effort, fault counts and the degree of testing. The most known quality models are as follows:

McCall's Quality Model

Concept

McCall is proposed in 1977 and aims at considering software quality factors from user and developers perspectives (Al-Qutaish 2010; Fenton and Bieman 2014).

Structure

McCall model presented in Figure 1 is structured into three main levels which are 11 quality factors that describe the user's view, 23 criteria to reflect the developer's perspective and 41 metrics to measure these criteria (McCall et al. 1977).

Figure 1. McCall quality model
Source: Authors' work

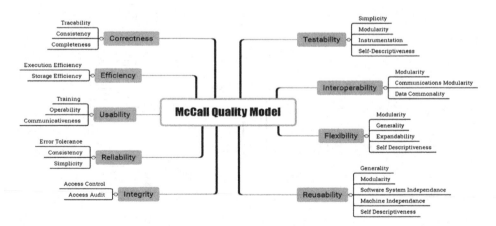

Benefits and Drawbacks

One of this model advantages is to identify the appropriate metrics for the different characteristics. It involves most of quality attributes and considers both users' and developers' perspectives. However, metrics used in this model are not clearly identified. Moreover, the functionality characteristic is not supported and hardware characteristics are not considered (Ortega 2003).

Boehm's Quality Model

Concept

It is proposed by Barry Boehm in 1978 to provide quantitative measurement of software quality factors and sub factors (Boehm et al. 1978; Al-Qutaish 2010).

Structure

Boehm model is defined through three levels of characteristics: the higher level is composed of 3 characteristics, the intermediate one consists of 7 characteristics and the lowest one is composed of 15 characteristics as presented in Figure 2 (Fenton and Bieman 2014). Concerning metrics, they may be defined from the primitive characteristics to measure them (Al-Qutaish 2010).

Figure 2. Boehm quality model
Source: Authors' work

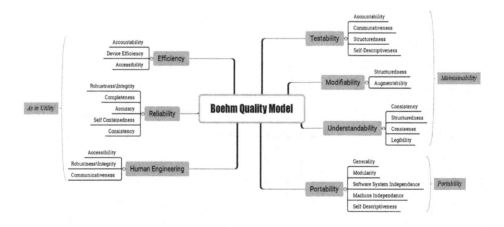

Benefits and Drawbacks

Boehm adds other characteristics to the Boehm model like human engineering and understandability. It includes the user needs and the developer perspectives since it supports maintainability characteristic (Ortega 2003). However, this model does not identify how to measure its quality characteristics (Al-Qutaish 2010).

FURPS Quality Model

Concept

It is suggested in 1987 by Hewlett Packard and Robert Grady. Its name is made through five basic characteristics which are functionality, usability, reliability, performance, and supportability (Ortega 2003).

Structure

FURPS model consists of two groups of requirements. The first group is functional requirements which includes capability, security and reusability. The second group is non-functional requirements which includes reliability, usability, supportability and performance as it is shown in Figure 3 (Al-Qutaish 2010).

Benefits and Drawbacks

The FURPS model includes the performance attribute that is not mentioned in McCall and Boehm models. However, it considers the software only from user's perspectives since it supports functionality but it does not support maintainability and reusability features (Ortega 2003).

Figure 3. FURPS quality model
Source: Authors' work

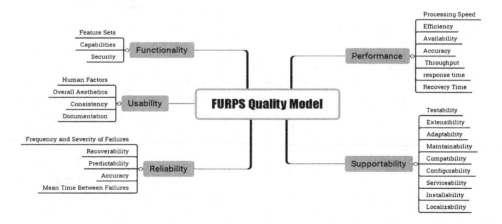

ISO 9126 and ISO 25010 Quality Models

Concept

The ISO 9126 Quality model was published by the ISO in 1991 as the standard for quality characteristics terminology. Then, in 2001, an expanded version was published, it contains the ISO quality models (ISO/IEC 9126-1 2003) and the proposed measures for them (Al-Qutaish 2010). After that, in 2011, the ISO/IEC 25010 replaces ISO/IEC 9126-1 (ISO/IEC 25010, 2011).

Structure

Two different models are proposed and classified into specific objectives: one of them is linked to the use of the software, whereas, the other is specific to the software production. We will focus here only on the latter in which the ISO 25010 product quality model which is based on eight basic characteristics as presented in Figure 4 (ISO/IEC 25010 2011). This model adds two characteristics to the ISO 9126-1 (ISO/IEC 9126-1 2003) which are security and compatibility.

Benefits and Drawbacks

ISO 25010 takes into account the security characteristic and it distinguishes between compatibility and portability characteristics which make it more relevant. However, similar to the ISO 9126-1, a detailed measures description is left and characteristics are not linked to metrics (Fenton and Bieman 2014).

Figure 4. ISO 25010 quality model
Source: Authors' work

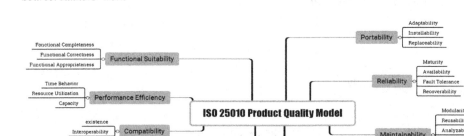

To sum up, authors deduce two major ideas. Different quality attributes are supported by at least one of the quality models. However, reliability is the only attribute that is supported by all of these models. Additionally, reliability was identified as an important attribute (Farooq et al. 2012). It has a much bigger concern in software engineering than the others because it allows the delivery of a failure-free software system. Measurement of software reliability will be the objective of the following sections. Second, the ISO 25010 is the only model that support security characteristic that is a very important software quality characteristic which make this model more relevant.

FUNDAMENTALS OF SOFTWARE MEASUREMENT

This section presents the basic elements of the software quality measurement.

Definition

Fenton defines measurement as the process of numbers assignment to entities' attributes in order to describe them with respect to a set of rules (Fenton 1994). Hence, measurement consists on collecting different information related to an entity which are required to quantitatively assess and describe it. This description is made by identifying characteristics using symbols or numbers related to this entity to distinguish it from others and to reflect the real world. For example, the statement "Student A is more intelligent than student B" is a measurement of students' degree of intelligence. However, to be more precise and to quantitatively measure this criterion, we can say: "the marks of student A are better than the marks of student B". In this example, the entity is the student, the measurement is the degree of intelligence of the student and the attribute is the student's marks.

Terminologies

It is undeniable to present software measurement regardless speaking about other related terminologies like entity, product, process, resources, metric, measure, etc as defined below:

- **Software Entities:** An object or an event in the real world can be considered as an entity which includes process, product and resources (Fenton and Pfleeger 1996).

- **Software Product:** Concerns all things which are related to the product that will be delivered to the user and include documentation, data, programs, etc (IEEE 610.12 1990; ISO/IEC 25010 2011).
- **Software Process:** Consists on the different activities which take place over its development like its duration and the effort associated to it (Bush and Fenton 1990).
- **Software Resources:** The items required by a process as inputs like personnel (individual or teams), materials, tools and methods (Bush and Fenton 1990).
- **Measure:** The number of symbol assigned to an entity and aims to characterize an attribute (Fenton and Bieman 2014). An example of measures used to assess attributes of an entity is presented in table 1.
- **Metric:** A quantitative measure of the different software quality attributes (IEEE 610.12, 1990; Fenton and Pfleeger 1996).
- **Measurement, Metric and Measure Relationship:** From the above definitions of measure, metric and measurement, we state that they have approximately the same meaning. Thus, we try to define a relation between them to help understand and distinguish these terms as presented in Figure 5.

Table 1. Example of entity and correspondent attributes and measures

Entity	Attribute	Measure
The program' code	Length of the program	The number of lines of code (LOC)
The program' code	Reliability	The Mean time to failure (MTTF) in hours of CPU.
The completed project	The duration of the project	The number of months from the start of the project to its delivery.

Source: (Fenton and Bieman, 2014)

Figure 5. Measure, metric and measurement relationship
Source: Authors' work

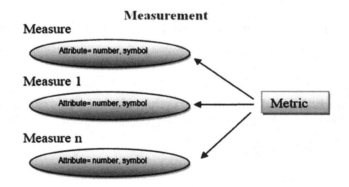

As it is shown above, a measure is defined in terms of an attribute, it characterizes an attribute by assigning to it a symbol or a number. A metric is derived from one or more measures. The whole procedure may refer to measurement.

Objective of Software Measurement

For Fenton and Bieman, measurement provides needed information for both managers and developers (Fenton and Bieman 2014):

- **For Managers:** Measurement helps them to measure the cost of the total project as well as the needed effort in the requirement, design, coding and testing steps. Added to that, they could identify the staff productivity in order to estimate the needed cost when changes are proposed. All in all, the quality of the project is based on these measurements that allow managers to compare it to other projects.
- **For Developers:** They can verify if requirements are quantitatively expressed by analyzing them. Furthermore, they are required to measure the number of faults in the different software phases (specification, code and test).

Concerning measurement uses, two major uses are discussed in literature, which are measurement for assessment and measurement for prediction (Fenton 1994):

- **Assessment Measurement:** Uses the existed software data to evaluate the current situation of the resources, the process and the developed product.
- **Predictive Measurement:** Identifies the relation between an attribute A to some existing measures of attributes A1… An. It aims at predicting the elements needed for future projects such as resources and time.

MEASUREMENT BY SOFTWARE METRICS

One of the famous statements related to this topic assumes that: "Anything that you need to quantify can be measured in some way that is superior to not measuring it at all" (Gilb 1988).

Need for Software Metrics

In literature, many approaches and tools are defined for software process improvement like Six Sigma, Goal Question-Metric and Capability Maturity Model have. Moreover, simple data analysis tools like Fishbone Diagram and Run Chart others are also proposed. For Fenton and Bieman (Fenton and Bieman 2014), measurement that are based on graphs and charts help customers and developers to decide if a project is successful. They note that measurements must be defined to identify for developers and customers how the project is made. One of these ways is software metrics which are so important that they achieve such goals:

- Collect data related to software resources, process and product to understand and control the software more easy as well as to identify which attribute affects the software and how it is influenced by it (Saini et al. 2014).
- Early detect which sections of the code have a low quality. Hence, the software' development and maintenance cost will be reduced (Stein et al. 2004). Moreover, Rosenberg et al. show that early detect and correct errors will help to prevent them later in the life cycle (Rosenberg et al. 1998).
- They are required to evaluate the software' performance and control the changes to it (Farooq et al. 2012).
- They help to measure, control, evaluate and improve the software process and product (Fenton and Bieman 2014).

As a conclusion, the use of software metrics can make much better estimations on the cost, length, effort and delivery time of a project which give them more accurate view of the overall project that realize a better project quality which eases and satisfies the customer.

Taxonomy of Software Metrics

There exist various classifications of software metrics based on various focuses as software entities, program paradigm and the way of computing these metrics. Starting with software entities, metrics are divided into three categories: process, project and product metrics (Saini et al. 2014). Considering the program paradigm, they may be also classified into procedural and object oriented metrics (Li 1998). Concerning the computing manner, metrics are divided into syntactic and semantic categories (Stein et al. 2004; Marsit et al. 2017). Thanks to these categories, software quality attributes (internal and external) can be reached.

Process, Project and Product Metrics

- **Process Metrics:** Measure all characteristics needed for the software development like needed time, effort, staff members, etc (Lyu 1996).
- **Product Metrics:** May include at the same time classical metrics like size and quality metrics as reliability (Mills 1988).

- **Project (Resources) Metrics:** Measures related to software project or resources including magnitude (the number of staff members), cost (the price of testing tools), and their quality (designers' experiences) (Fenton and Bieman 2014).

Procedural and Object Oriented Metrics

- **Procedural Metrics:** Defined in procedural paradigm like Halstead and McCabe Cyclomatic complexity metrics (Brian et al. 1996).
- **Object Oriented Metrics:** Proposed to overcome shortcomings of procedural metrics and to measure different OO concepts i.e cohesion, inheritance, encapsulation, etc (Li 1998).

Syntactic and Semantic Metrics

This classification includes the same metrics of the previous classifications, but it is viewed from another perspective: way of computing these metrics.

- **Syntactic Metrics:** Focus on the syntax of the program's code to estimate some quality attributes like size and complexity (Mili et al. 2014; Amara and Rabai 2017).
- **Semantic Metrics:** Reflect what a program does, and require understanding the functionality of the code (Stein et al. 2014; Mili et al. 2014; Marsit et al. 2017).

In spite of the different manner of computing, syntactic and semantic metrics are complementary since the use of software syntactic and semantic information may lead to an effective measurement of software attributes.

There are a lot of semantic metrics mentioned in literature which will be presented in details in the following chapter. This chapter will focus only on the syntactic metrics.

SYNTACTIC METRICS

They generally include traditional metrics and OO metrics.

Procedural Metrics

The most discussed traditional metrics are as follows:

Line of Code (LOC)

Concept

LOC is the program text lines like program headers, declarations, executable and no executable statements regardless comments or blank lines (Conte et al. 1986).

Criticism

The treatment of black lines and comments is ambiguous which makes LOC' attitude misleading.

Function Points (FP)

Concept

Proposed by Albrecht in 1979 to estimate the size of the program from the user's requirements before writing the code. FP can be also applied lately to estimate the size and therefore the cost and productivity (Albrecht et al. 1983).

Criticism

LOC and FP provide developers and engineers with an idea about the size of the program. However, FP may be more useful for predicting the work effort than LOC since they can be computed early from requirements in the development process (Fenton and Bieman 2014).

Halstead's Product Metrics

Concept

These are introduced in 1977 to measure the complexity of the program through other attributes like the length, the level, the value and the programming effort using the operators and the total number of operands (variables or constants) (Halstead 1977).

Criticism

Generally, this set of metrics is proposed not only to measure the size and other program aspects but also to measure its complexity and defect rates. However, the distinction between operands and operators is hard which creates a difficulty in the computing process (Mills 1988).

McCabe's Cyclomatic Complexity (CC)

Concept

CC is developed by Thomas McCabe in 1976 (McCabe 1976). It is used as an indicator of the program' complexity by counting the number of independent paths in the program' graph (Mills 1988).

Criticism

A low value of CC decreases the risk to modify the program. It is hard to distinguish between different conditional statements. Moreover, this metric highlights the control graph complexity (Saini et al. 2014).

In short, presented procedural metrics are so used in estimating basic quality attributes such as size and complexity which are used to estimate other attributes like cost, effort and productivity. Authors summarize the procedural metrics and their use throughout the development process in Table 2.

Object Oriented Metrics

Different suites of OO metrics have been presented in literature (Radjenović et al. 2013). Most of them are based on the ISO model and presented as follows:

Chidamber and Kemerer (C&K) OO Suite

C&K suite is suggested in 1991. It consists of six metrics applied for OO software design (Chidamber and Kemerer 1994). This suite is considered as the basic of OO software design metrics (Kumar and Kaur 2011). To make an emphasis on that, we present below a detailed description of this suite and we give a general description of the others.

Table 2. Procedural metrics and their use through SDLC

Metric	Development Phases		
	Requirement	Design	Code
LOC			*
FP	*		*
Halstead			*
McCabe Cyclomatic Complexity			*

Source: Authors' work

- **Weighted Method per Class (WMC):**
  - **Concept:** It is used to compute the class using the sum of all its methods' complexity (Etzkorm Delugach 2000). It measures the complexity and helps to identify the time and effort required for the development of this class (Kumar and Kaur 2011).
  - **Criticism:** WMC is a good complexity measurement metric for its validation of the Weyker's analytical properties except the sixth one (Chidamber and Kemerer 1998; Weyuker 1988). However, its interpretation is difficult since the number of methods and their complexities are two independent attributes (Kumar and Kaur 2011).
- **Depth of Inheritance Tree (DIT):**
  - **Concept:** DIT measures the inheritance' depth in the class, it identifies the different classes affecting this class through inheritance (Chidamber and Kemerer 1994).
  - **Criticism:** For Li, the difference between the maximum length from the node to the root and the number of classes affecting the current class through inheritance make the DIT definition ambiguous and inappropriate (Li 1998). Moreover the maximum length between the root and the node is difficult to calculate because we can have different roots (Kumar and Kaur 2011).
- **Number of Children (NOC):**
  - **Concept:** The objective is to count the class' direct children i.e direct subclasses or descendant that inherit the methods of the class (Briand et al. 2000).
  - **Criticism:** NOC meets most of metrics validation' properties of Weyuker's except the sixth one (Weyker 1988). If the NOC value increases, then, the number of required testing methods increases too. Hence, the class complexity will also increase. (Chidamber and Kemerer 1994). Besides, NOC seems to be insufficient for counting subclasses since one class has an effect on its subclasses, not only its direct subclasses (Etzkorn and Delugach 2000).
- **Lack of Cohesion of Methods (LCOM):**
  - **Concept:** Mathematically, LCOM is the number of disjoint methods (have no common instance variable) minus those who have one common instance at least (Etzkorn and Delugach 2000).
  - **Criticism:** The more the LCOM value is closer to 0, the more the cohesion is better which promotes lower complexity and increases reliability (Vipin Kumar et al. 2011). LCOM fails to satisfy the property 4 and 6 of Weyuker (Chidamber and Kemerer 1994; Weyuker 1988).

- **Coupling Between Objects (CBO):**
    - ○ **Concept:** CBO of a class consists on counting the number of other classes coupled to it. A class is coupled to another one if the methods of the first class are used by the other one and vice versa (Briand et al. 1996).
    - ○ **Criticism:** Higher coupling leads to more complex tests (Kumar et al. 2011). Besides, CBO fails to satisfy the sixth Weyuker's property (Weyker 1988).
- **Response for a class (RFC):**
    - ○ **Concept:** RFC is a measure of class' coupling, it counts the methods which are executed as a response to a message received from an object to the current class (Kumar and Kaur 2011).
    - ○ **Criticism:** If the number of methods used to respond the received message, then more testing and debugging will be required, this will directly increase the complexity. The RFC fails to satisfy the sixth Weyuker's property (Kumar and Kaur 2011; Weyuker 1988).

Li and Henry OO Metrics

In 1993, Li and Henry proposed a set of five OO metrics whose aim is to measure object coupling through three dimensions of coupling which are (1) inheritance, (2) abstract data type and (3) message passing:

- **Message Passing Coupling (MPC):** Proposed to measure the coupling between classes using messages sent from an object to another. A large value of MPC increases the class complexity (Li and Henry 1993; Briand et al. 2000).
- **Data Abstraction Coupling (DAC):** The number of abstract data type (ADT). A large value of DAC increases the coupling between that class and other classes (Briand et al., 2000).
- **Number of Methods (NOM):** Measures the number of local methods in the class that may indicate the operation property in the class (Li and Henry 1993).
- **Size1 and Size 2:** Size 1 measures the number of semicolons in a class. Size 2 measures the number of attributes and local methods in the class (Li and Henry 1993).

Abreu et al. OO MOOD Metrics

In 1995, a set of six OO metrics called MOOD (Metrics for Object-Oriented Design) are propose to measure different OO mechanisms including encapsulation and polymorphism (Harrisson 1998):

- **Method Hiding Factor and Attribute Hiding Factor (MHF and AHF):** MHF measures the hidden information based on the visibility of the methods. Whereas, MHF uses measures the hidden information based on the visibility of the attributes (Harrisson 1998).
- **Method Inheritance Factor and Attribute Inheritance Factor (MIF and AIF):** These metrics are proposed to measure the direct number of attributes and inherited methods which are a proportion of the total number of attributes and methods (Harrisson 1998; Chawla 2012).
- **Coupling Factor (CF):** A direct measure of the coupling between classes described through the two classes' relationship. Higher value of CF drows higher level of interclass coupling (Harrisson 1998; Chawla 2012).
- **Polymorphism Factor (PF):** Proposed to measure the polymorphism property (Harrisson 1998; Chawla 2012).

Li OO Metrics

In 1998, a set of six OO metrics are proposed as a remedy of C&K suite' deficiencies. The Li suite's purpose is to measure different OO concepts such as coupling and inheritance (Li 1998; Kumar and Kaur 2011).

- **Number of Ancestor Classes (NAC):** Similar to DIT metric and counts the total number of a class' ancestor classes based on inheritance (Li 1998; Kumar and Kaur 2011).
- **Class Method Complexity (CMC):** Similar to WMC and consists on measuring the local methods' complexity. The large value of value of CMC increases the complexity of the class (Li 1998; Kumar and Kaur 2011).
- **Coupling Through Abstract Data Type (CTA):** A measure of a class' coupling using classes that are based on abstract data types and coupled to this class. A large value of CTA is beneficial for test engineer to design test cases as well as understand the class for the maintenance engineer (Li 1998; Kumar and Kaur 2011).
- **Coupling Through Message Passing (CTM):** Proposed to measure coupling through the number of different messages sent out from one class to other

classes. A large value of CTM increases the time and effort needed by a test engineer to design test cases. Moreover, more specific methods are needed to help maintenance engineer to understand solve problems (Li 1998; Kumar and Kaur 2011).

- **Number of Descendant Classes (NDC):** Similar to NOC and aims at measuring the subclasses number affected by a class and not only its immediate subclasses (Li 1998; Kumar and Kaur 2011).
- **Number of Local Methods (NLM):** Proposed to measure the number of local methods which may be accessible outside the class on which they are defined. The required effort to the software design or implementation is increased by the large value of NLM (Li 1998; Kumar and Kaur 2011).

Bansiya and Davis OO QMOOD Metrics

In 2002, 11 OO metrics called QMOOD (Quality model for object-oriented design) are proposed (Bansiya and Davis 2002). These metrics are proposed not only to measure OO i.e inheritance and polymorphism but also to empirically study their correlation with different attributes like functionality, flexibility, reusability and effectiveness (Chawla 2012).

- **Design Size in classes (DSC):** Measures the total number of design' classes. It is used to study its correlation with the quality attributes. A high value of DSC increases the functionality and reusability of the program (Bansiya and Davis 2002).
- **Number of Hierarchies (NOH):** Measures the design hierarchies using the number of classes. As NOH goes up, the class functionality increases (Bansiya and Davis 2002).
- **Average Number of Ancestors (ANA):** Measures the average number of a class' ancestors. The abstraction level that has a negative influence on the understandability and a positive effect on the extendibility and effectiveness attributes (Bansiya and Davis 2002).
- **Data Access Metric (DAM):** Divides the number of private attributes by the total number of declared attributes in order to assess the encapsulation property of this class. A large value of this metric influences positively on other attributes like understandability, flexibility, understandability and effectiveness attributes (Bansiya and Davis 2002).
- **Direct Class Coupling (DCC):** Measures the coupling OO property by identifying different classes on which a class is directly related. A high value of DCC has a negative influence on reusability and extendibility attributes (Bansiya and Davis 2002).

- **Measure of Aggregation (MOA):** Measures the composition property through user defined data declaration. A high value of MOA has a positive influence on the flexibility (Bansiya and Davis, 2002).
- **Cohesion Among Methods of Class (CAM):** Measures the cohesion property and computes the link between using the list of methods parameter. High value of CAM means high level of cohesion and has a positive influence on the reusability, understandability and functionality (Bansiya and Davis 2002).
- **Measure of Functional Abstraction (MFA):** Divides the number of the inherited methods by the total number of methods in order to measure the class' inheritance. High value of MFA has a positive influence on the effectiveness attribute (Bansiya and Davis 2002).
- **Number of Methods (NOM):** Counts the total number of methods in a class in order to assess its complexity. The complexity property has a negative influence on the understandability attribute (Bansiya and Davis 2002).

Sum Up of OO Syntactic Metrics

To highlight the importance of presented OO syntactic metrics, authors present in Table 3 a sum up of literature review of the well-known related studies focused on these metrics.

Previous syntactic metrics prove their effectiveness for software quality assessment. In summary, authors deduced the following notes:

- By analyzing shortcomings related to different OO metrics suites, authors found that C&K suite is the most referenced one for several reasons. First of all, most of these metrics are computed at the class level that is the basic unit of OO paradigm by contrast to the other suites basically computed for methods and attributes level. Moreover, this suite covers essential OO concepts like inheritance, coupling, cohesion, etc. Besides, most of the other suites are proposed as alternatives to C&K metrics, whereas, they have the same objective and the same theoretical basis.
- Most of OO metrics are proposed as design metrics, however, we stated that all of them are computed at the code level and require a fully implementation of the program. Hence, we cannot consider syntactic metrics as design metrics.
- Generally OO metrics are used for assessing internal attributes like cohesion, coupling and inheritance; however, quality of software systems is described through internal and external attributes. Regarding the application of these metrics, we can note that they still insufficient in measuring external quality attributes including reliability.

Table 3. Literature review of the well-defined procedural and OO syntactic metrics

Study	Metrics	Development Phase	Property to Measure	Metric' Correlation With Software Quality Attributes
Procedural Syntactic Metrics				
Conte et al. 1986.	LOC	Coding	Program size	LOC reflects the program complexity
Albrecht et al. 1983	FP	Requirement and code	Program size,	FP reflects the cost and the productivity
Halstead 1977	Halstead	Coding	Complexity	Used as an indicator of the program complexity (Mills 1988).
McCabe 1976.	CC	Coding	Complexity	• CC reflects the program complexity. • CC reflects the program flexibility and maintainability: a low value of CC decreases the risk to modify the program.
OO Syntactic Metrics				
Chidamber and Kemerer 1994	WMC, DIT, NOC, LCOM, RFC, CBO	Design, code	Complexity, Inheritance Length, cohesion, coupling	• The larger the WMC and DIT are used for fault detection; the larger the DIT and WMC, the larger the probability of fault detection. • LCOM is used as predictor of fault-prone classes (Basili 1996). • RFC and CBO are used as predictors of number of faults and the maintenance effort (Abreu and Melo, 1996)
Abreu et al. 1996	MIF, AIF, MFA, MHF, AHF, CF,	Design, code	Inheritance Length, Encapsulation, Coupling	• MFA, MIF and AIF predict the maintenance effort and the number of faults (Briand et al. 2000; Basili et al. 1996) • MHF and AHF have a significant correlation with defect density and normalized rework (Abreu and Melo, 1996)
Li 1998	DAC, DCC, CF	Design and code	Coupling	Predict the maintenance effort and the number of faults (Abreu and Melo, 1996).
Bansiya and Davis 2002	CAM, DAM, NOP	Design and code	Cohesion, encapsulation, polymorphism	• CAM is used as predictor of fault-prone classes (Basili et al. 1996) • DAM has a significant correlation with defect density and normalized rework (Abreu and Melo, 1996). • NOP has a significant influence on various quality attributes like reusability (Bansiya et al. 2012).

Source: Authors' work

- Syntactic metrics still insufficient compared to semantic ones which are computed from early development process phases and can measure directly external quality attributes. However, we can investigate them to other levels of software assessment. For example, they may predict the maintenance effort as presented above. Moreover, they can be used to make a comparative study of the different OO languages like Java, C sharp and C++ languages. This comparison may help developers to select the appropriate language to build their OO systems.
- Syntactic metrics are important since we can investigate them to answer the following question: How OO syntactic metrics may influence external quality characteristics such as reliability through defect density (error and

fault density)? To answer this question, we propose in the following section, a qualitative approach in which we will identify how we can measure the software reliability based on the correlation that may be exist with these metrics.

Syntactic Software Tools

To collect the different metrics from the program representation, appropriate tools are required. Lincke et al. present the well-developed ones like Analyst 4j, Chidamber & Kemerer Java Metrics, CCCC, Eclipse OO Meter, Semmle, etc (Lincke et al. 2008). Table 4 gives a comparison between these tools based on used language and metrics computation criteria.

The presented metrics depend on the tool to be used. Hence, using different tools to compute a same metric with the same input may give different results which make the metrics' validation a difficult task.

SOFTWARE RELIABILITY MEASUREMENT

As reliability is one of software dependability attributes used to reflect the user satisfaction, however its measurement is difficult since there are no lucid definitions of its various aspects (Vipin Kumar 2011).

Table 4. Syntactic metrics tools

Tools	Accepted Source Code Files	Metrics
Analyst4j	Java	Measures various metrics
CCCC	Java, C++	Measures various metrics
Chidamber & Kemerer Java Metrics	Java	C&K metrics
Eclipse OOMeter	Java	Measures various metrics
Understand for Java	Java	Measures various metrics
OOMeter	Java, C#	Measures various metrics
Resource Standard Metrics (RSM)	C, C++, C# and Java	Measures various metrics
QMOOD++	C++	QMOOD metrics

Source: Authors' work

Terminologies

In fact, authors can't speak about reliability measurement without integrating fault, error, failure terminologies:

- **Fault:** A mistake in a program that may be caused by a programmer and cause a failure (Lyu 1996).
- **Error:** Can occur when an element of the computer software produce unwanted state (Lyu 1996).
- **Failure:** Occurs when the program operation deviates from program needs which means that the final program does not meet the required specifications (Chidamber and Kemerer 1994).

Ways of Software Reliability Measurement

Generally reliability measurement involves reliability estimation and prediction (Lyu 1996; Fenton and Bieman 2014).

- **Reliability Estimation:** Consists on using statistical techniques on failure data to assess past and current reliabilities (IEEE 1633 2008).
- **Reliability Prediction:** Aims at using existing software failure data and other related attributes from the beginning of SDLC to predict the values of other attributes and future reliability (Lyu 1996).

There are different ways of software reliability measurement presented in literature which covers reliability techniques, models and metrics as presented under here.

Software Reliability Models (SRM)

SRM has originated since the early of 1970 and consists on using statistical approaches in the later phases of software development (testing and operation). The objective is to show the relationship between the failures' process and the different quality attributes (Goel 1985; Lyu 1996). This can help estimating the software reliability using the rate of failures or the time between them (Fenton 1994; Farooq et al. 2012; Amara and Rabai 2017).

Numerous software reliability models are presented in literature as Jelinski-Moranda, Musa, and Nonhomogeneous Poisson Process (NHPP) models. Generally, these models are classified into two categories; time between failures and fault-count models (Lyu 2007; Fenton and Pfleeger 1996).

- **Time Between Failure Models:** These models use the mean time between failures in a specified period of time to estimate the expected time to the next fault (Farooq et al. 2012).
- **Fault-Count Models:** Its objective is to model the number of faults detected in a specific time interval for instance the CPU execution time (Goel 1985).

To support the reliability database and to automate the data collection, different tools are proposed to estimate these models' parameters (IEEE 1633 2008). Most of them are Statistical Modeling and Estimation of Reliability Functions for Software (SMERFS) from the Naval Surface Warfare Center (NSWC/DD) and, Computer Aided Software Reliability Estimation (CASRE) from the Open Channel Foundation (Goel 1985).

Proposed reliability models help to predict the software failure behavior. The objective is to identify the corrective measures to minimize the failure rate or to increase the time between failures. A good model should not be limited to failure description, but to be able to predict these failures (Mills 1998).

To sum up, most of these models are used during test and operation to estimate reliability and not to predict it (Goel 1985).

Software Reliability Techniques

These techniques are needed to minimize the occurrence of faults. four major classes of these techniques are defined:

- **Fault Prevention:** The initial defensive mechanism used to avoid fault occurrence in the system to ameliorate its reliability and include formal methods (Statecharts, Petri Nets) and software reuse (Valido-Cabrera 2006).
- **Fault Removal:** All software faults cannot be avoided using fault prevention techniques. Fault removal techniques are then proposed to identify and eliminate them (Lyu 1996; Lyu 2007). They include software testing and formal inspection that is carried out by small groups of peers to find faults, correct them and verify them (Lyu 2007).
- **Fault Tolerance:** Aims at continuing the execution of the program despite the presence of errors (Jaoua and Mili 1990). They may be singe-version techniques like subdividing the program on different modules, checkpointings and restarting or multiple-versions techniques like N-version programming (Lyu 2007).
- **Fault Forecasting:** The objective is to estimate and predict occured failures by studying the operational environment and developing reliability models (Lyu 1996).

Software Reliability Metrics

Software metrics are proposed to measure quality attributes in testing phase but may be also used to predict other external attributes based on internal ones. They may be classified into two direct and derived metrics.

- **Direct Metrics for Reliability Assessment - Mean Time To Failure (MTTF):** It measures an attribute without depending on other measures. For instance, reliability is quantified using the mean time to failure (MTTF) that is the operation time until the next system failure (IEEE 1061 2009; Fenton and Pfleeger 1996).
- **Derived Metrics for Reliability Assessment:** These metrics are derived from the direct ones (Fenton and Bieman 2014). Presented metrics are succeeding in measuring internal attributes directly from source code but still limited for measuring external ones. These attributes can be only indirectly measured from internal attributes that may provide important indicators of external ones (Fenton and Pfleeger 1996).

Since this study focuses on measuring software reliability, authors may use the complexity which as an important attribute measured from the source code to identify how reliability may be evaluated. For that, Table 5 shows the optimal value

Table 5. Preferred values of OO metrics and their impact on complexity

Metrics	Optimal Values and Impact on Complexity
WMC	The preferred WMC' value should be between 100 and 200. A higher value of WMC increases complexity (Rosenberg et al., 1998).
DIT	DIT should be less than or equal to 4. A higher value of DIT in the class increases its complexity (Kumar and Kaur, 2011).
NOC	Lesser the better. A higher value of NOC increases the complexity of the class.
LCOM	LCOM should be closer to 0 (Kumar and Kaur, 2011). A higher value of LCOM increases the complexity of the class (Valido-Cabrera, 2006).
CBO	CBO should be in the interval [0-4]. A higher value of CBO increases the complexity (Kumar and Kaur, 2011).
RFC	Lesser the better. A higher value of RFC increases the complexity of the class (Valido-Cabrera, 2006).
MPC	Lesser the better. A higher value of MPC increases coupling. Hence complexity of the class also increases (Li and Henry, 1993).
DAC	Lesser the better. A higher value of DAC increases coupling. Hence complexity of the class also increases (Li and Henry, 1993).
CF	Lesser the better. A higher value of CF indicates higher level of interclass coupling which increases the complexity (Saini et al., 2014).
CMC	Lesser the better. If the CMC is large, then the class complexity will be large too (Li, 1998).

Source: Authors' work

of various metrics presented in this chapter, and the relationship between attributes measured by these metrics and complexity.

To identify the correlation between these metrics and reliability, different statistical tools like Stata or R may be used to verify the hypothesis "OO metrics have a strong impact on reliability". In order to interpret this relationship, two variables have to be considered which are the correlation coefficient and the probability levels (p-value):

- The correlation coefficient helps to identify if there is a relationship between a metric and a quality attribute like reliability; this correlation may be positive, negative or not significant which means that the metric will not have any impact on reliability.
- The p-value measures the significance of the relationship. A small value of the p-level (<0.05) means that the relationship is significant.

The correlation between OO metrics and reliability can identify which of these metrics can be considered as the sole major predictors of reliability.

FUTURE RESEARCH DIRECTIONS

Basing on what has been presented, authors aim at empirically validate the effectiveness of OO syntactic metrics for estimating reliability. To this end, developing a compiler writing tool in order to obtain optimal values for further metrics is needed. Literature shows that current trends on software measurement are focusing on semantic metrics, the following chapter will present most of semantic metrics proposed for software quality assessment. Added to that, it will describe how these metrics are empirically validated. Concerning no validating ones, an empirical validation will be proposed to ensure the correlation between semantic metrics and quality attributes like defect density which leads researchers to get a better view of reliability prediction.

CONCLUSION

Software measurement is not a new concept. However with the rapid evolution in technology and the development of more complex software systems, more studies and continuous investigations in this field still needed. In this chapter, we presented most elements that are related to the measurement concept in the software engineering context.

In the first part, we discussed the importance of quality in software engineering by presenting most important software quality models that subdivide quality into factors or attributes and show their relationships. Two main classes of these attributes are defined; internal and external attributes. A comparison between quality models shows that reliability is one the most important software quality factor.

In the second part, we discussed fundamentals of software measurement based on software metrics which are divided into two types which are syntactic and semantic. Syntactic metrics are important to predict external attributes based on internal ones.

The presented chapter is benefit: it studies fundamentals of software measurement and gives clear definitions of related terminologies to avoid ambiguities. Added to that, it shows how internal attributes can be measured through the development process and how they may be used as predictors to estimate external ones.

REFERENCES

Abreu, F. B., Goulao, M., & Estevers, R. (1995). Toward the Design Quality Evaluation of OO Software Systems. *Proc. Fifth Int'l Conf. Software Quality.*

Abreu, F. B., & Melo, W. (1996). Evaluating the Impact of OO Design on Software Quality. *Proc. Third Int'l Software Metrics Symp.* 10.1109/METRIC.1996.492446

Al-Qutaish, R. E. (2010). Quality Models in Software Engineering Literature: An Analytical and Comparative Study. *The Journal of American Science, 6*(3), 166–175.

Albrecht, A. J., John, E., & Gaffney, J. R. (1983). Software Function, Source Lines of Code, and Development Effort Prediction: A Software Science Validation. *IEEE Transactions on Software Engineering, SE-9*(6), 639–648. doi:10.1109/TSE.1983.235271

Amara, D., & Rabai, L. B. A. (2017). Towards a new framework of software reliability measurement based on software metrics. *Procedia Computer Science, 109*, 725–730. doi:10.1016/j.procs.2017.05.428

Bansiya, J., & Davis, C. G. (2002). A hierarchical model for object-oriented design quality assessment. *IEEE Transactions on Software Engineering, 28*(1), 4–17. doi:10.1109/32.979986

Basili, V. R., Briand, L. C., & Melo, W. L. (1996). A validation of object-oriented design metrics as quality indicators. *IEEE Transactions on Software Engineering, 22*(10), 751–761. doi:10.1109/32.544352

Bevan, N. (1999). Quality in use: Meeting user needs for quality. *Journal of Systems and Software, 49*(1), 89–96. doi:10.1016/S0164-1212(99)00070-9

Boehm, B. W., Brown, J. R., Kaspar, J. R., Lipow, M., & MacCleod, G. (1978). *Characteristics of Software Quality*. Elsevier Science Ltd.

Briand, L. C., Morasca, S., & Basili, V. R. (1996). Property-based software engineering measurement. *IEEE Transactions on Software Engineering*, 22(1), 68–86. doi:10.1109/32.481535

Briand, L. C., Wüst, J., Daly, J. W., & Porter, D. V. (2000). Exploring the relationship between design measures and software quality in object-oriented systems. *Journal of Systems and Software*, 51(3), 245–273. doi:10.1016/S0164-1212(99)00102-8

Bush, M. E., & Fenton, N. E. (1990). Software measurement: A conceptual framework. *Journal of Systems and Software*, 12(3), 223–231. doi:10.1016/0164-1212(90)90043-L

Chawla, S. (2012). Review of MOOD and QMOOD metric sets. *International Journal of Advanced Research in Computer Science and Software Engineering*, 3(3), 448–451.

Chhikara, A., Chhillar, R.S., & Khatri, S. (2011). *Applying Object Oriented Metrics to C#(C Sharp) programs*. Academic Press.

Chidamber, S. R., & Kemerer, C. F. (1994). A metrics suite for object oriented design. *IEEE Transactions on Software Engineering*, 20(6), 476–493. doi:10.1109/32.295895

Conte, S. D., Dunsmore, H. D., & Shen, V. Y. (1986). *Software Engineering Metrics and Models*. Menlo Park, CA: Benjamin-Cummings.

Etzkorm, L., & Delugach, H. (2000). Semantic Metrics Suite for Object-Oriented Design. *34th International Conference on Technology of Object-Oriented Languages and Systems*.

Farooq, S. U., Quadri, S. M. K., & Ahmad, N. (2012). Metrics, Models and Measurements in Software Reliability. *IEEE 10th International Symposium on Applied Machine Intelligence and Informatics (SAMI)*, 441 - 449.

Fenton, N. (1994). Software Measurement: A Necessary Scientific Basis. *IEEE Transactions on Software Engineering*, 20(3), 199–206. doi:10.1109/32.268921

Fenton, N., & Bieman, J. (2014). Software Metrics: A Rigorous and Practical Approach (3rd ed.). Academic Press. doi:10.1201/b17461

Fenton, N., & Pfleeger, S. L. (1996). *Software Metrics: A Rigorous & Practical Approach* (2nd ed.). International Thomson Computer Press.

Gilb, T. (1988). *Principles of Software Engineering Management* (2nd ed.). Reading, MA: Addison-Wesley.

Goel, A. L. (1985). Software reliability models: Assumptions, limitations and applicability. *IEEE Transactions on Software Engineering, 11*(12), 1411–1423. doi:10.1109/TSE.1985.232177

Grady, R. B., & Caswell, D. (1987). *Software Metrics: Establishing a Company-Wide Program*. Englewood Cliffs, NJ: Prentice-Hall.

Halstead, M. H. (1977). *Elements of Software Science (Operating and programming systems series)*. New York: Elsevier Science Inc.

Harrison, R., & Steve, J. (1998). An Evaluation of the MOOD Set of Object-Oriented Software Metrics. *IEEE Transactions on Software Engineering, 24*(6), 491–496. doi:10.1109/32.689404

Helali, R. G. M. (2015). A Comparison between Semantic and Syntactic Software Metrics. *International Journal of Advanced Research in Computer and Communication Engineering, 4*(2).

IEEE Standard 1061, Software Quality Metrics Methodology, IEEE Computer Society Press, New York, 2009.

IEEE Standard 610.12-1990, Glossary of Software Engineering Terminology, IEEE Computer Society Press, New York, 1990.

IEEE Std 1633 - IEEE Recommended Practice on Software Reliability, 2008.

International Standards Organisation, Software engineering—Product quality Part 1: Quality model, SS-ISO/IEC 9126-1, 2003.

International Standards Organization. (2011). Systems and software engineering—systems and software quality requirements and evaluation (SQUARE)—systems and software quality models, ISO/IEC 25010.

Jaoua, A., & Mili, A. (1990). The Use of Executable Assertions for Error Detection and Damage Assessment. *J. Software Systems, 12*, 15-37.

Kumar, R., & Kaur, G. (2011). Comparing Complexity in Accordance with Object Oriented Metrics. *International Journal of Computer Applications, 5*(8).

Li, W. (1998). Another metric suite for object-oriented programming. *Journal of Systems and Software, 44*(2), 155–162. doi:10.1016/S0164-1212(98)10052-3

Li, W., & Henry, S. (1993). Maintenance metrics for the Object-Oriented paradigm. *Proceedings of First International Software Metrics Symposium*, 52-60. 10.1109/METRIC.1993.263801

Lincke, R., Lundberg, J., & Löwe, W. (2008). Comparing software metrics tools. *Proceedings of the 2008 international symposium on Software testing and analysis*, 131–142.

Lyu, M. R. (1996). Handbook of Software Reliability Engineering. IEEE Computer Society Press.

Lyu, M.R. (2007). Software Reliability Engineering: A Roadmap. *Future of Software Engineering*.

Marsit, I., Omri, M. N., & Mili, A. (2017). *Estimating the Survival Rate of Mutants*. Academic Press.

McCall, J.A., Richards, P.K., & Walters, G.F. (1977). *Factors in software quality*. RADC TR-77-369, 1977. Vols I, II, III, US Rome Air Development center Reports NTIS AD/A-049 014, 015, 055.

Mili, A., Jaoua, A., Frias, A., & Helali, R. G. M. (2014). Semantic metrics for software products. *Innovations in Systems and Software Engineering, 10*(3), 203–217. doi:10.100711334-014-0233-3

Mili, A., & Tchier, F. (2015). Software Testing: Concepts and Operations. In Fundamentals of Software Engineering (2nd ed.). Academic Press.

Mills, E. (1998). *Software Metrics. SEI Curriculum Module SEI-CM-12-1.1*. Carnegie Mellon University, Software Engineering Institute.

Ortega, M., Pérez, M. Í., & Rojas, T. (2003). Construction of a Systemic Quality Model for Evaluating a Software Product. *Software Quality Journal, 11*(3), 219–242. doi:10.1023/A:1025166710988

Radjenović, D., Heričko, M., Torkar, R., & Živkovič, A. (2013). Software fault prediction metrics: A systematic literature review. *Information and Software Technology, 55*(8), 1397–1418. doi:10.1016/j.infsof.2013.02.009

Rosenberg, L., Hammer, T., & Shaw, J. (1998). Software metrics and Reliability. *9th International Symposium on Software Reliability*.

Saini, N., Kharwar, S., & Agrawal, A. (2014). A Study of Significant Software Metrics. *International Journal of Engineering Inventions, 3*.

Stein, C., Etzkorn, L., Cox, G., Farrington, Ph., Gholston, S., Utley, D., & Fortune, J. (2004). A New Suite of Metrics for Object-Oriented Software. *Proceedings of the 1st International Workshop on Software Audits and Metrics*, 49-58.

Valido-Cabrera, E. (2006). *Software reliability methods*. Technical University of Madrid.

Vipin Kumar, K. S., Albin, A. I., Arjunal, B., & Sheena, M. (2011). Compiler Based Approach To Estimate Software Reliability Using Metrics. *Annual IEEE India Conference*.

Weyuker, E. J. (1988). Evaluating Software Complexity Measures. *IEEE Transactions on Software Engineering, 14*(9), 1357–1365. doi:10.1109/32.6178

Wong, W.E., Debroy, V., & Restrepo, A. (2009). *The Role of Software in Recent Catastrophic Accidents*. IEEE Reliability Society 2009 Annual Technology Report.

Chapter 8
A Study on Risk Management in Financial Market

Smruti Rekha Das
SOA University, India

Kuhoo
College of Engineering and Technology, India

Debahuti Mishra
SOA University, India

Pradeep Kumar Mallick
Gurukula Kangri Vishwavidyalaya, India

ABSTRACT

The basic aim of risk management is to recognize, assess, and prioritize risk in order to assure that the uncertainty should not deviate from the intended purpose of the business goals. Risk can take place from various sources, which includes uncertainty in financial markets, recessions, inflation, interest rates, currency fluctuations, etc. Various methods used for this management of risk are faced with various decisions such as the market price, historical data, statistical methodologies, etc. For stock prices, the information derives from the historical data where the next price depends only upon the current price and some of the outside factors. Financial market is very risky to invest money, but the proper prediction with handling the risk will benefit a lot. Various types of risk in the financial market and the appropriate solutions to overcome the risk are analyzed in this study.

DOI: 10.4018/978-1-5225-5793-7.ch008

INTRODUCTION

Financial risk is basically a risk associated with financing, which includes financial transactions such as purchasing, loan, mortgage, bank account, credit card, debit card etc. In terms of handling the business in finance, the risk is faced; apart from managing the financial risk other types of risk may come, which has a great impact over the business phenomenon such as natural disasters, political issues, disease breakouts by giving an importance to employee health issues etc. Various types of risk is related with financial market such as market risk, liquidity risk, equity price risk, foreign exchange risk, commodity price risk, credit risk, foreign investment risk, operational risk, model risk, financial risk etc. The financial risk management concept dramatically changes on various future time horizons.

In this study, we will analyze about various types risk and how to overcome from the risk with the effective factor of the solution. This study undergoes the analysis of various types of risk occurring in the financial market. The work flow of the paper is ordered as follows: section 2 analyses over various risk factors of financial market, section 3 describes about approaches to solve the issues arising in the form of risk in the financial market and finally section 4 concludes the study.

ANALYSIS OF VARIOUS TYPES OF FINANCIAL RISK

Market Risk

Market risk arises due to the market price movements of financial instruments. Under market risk another sub category of risk is associated, which is known as interest risk. It arises due to the movement of interest rate. Through this the interest related price is affected, such as bonds, loans, etc. If the interest rate will be increased, then the value of bonds will be decreased. To manage this type of risk, various hedges are available, such as forward rate agreements and interest rate swaps. Hedging concept is introduced to mitigate the risk basically reduce any type of substantial losses, when a combination of assets is selected for investment to offset the movements. When investing in stocks, simultaneously there is a sell option is available to sell out that stock at some point in the future. Like hedging, derivatives are also used to reduce various types of risk. Market risk can be arranged in various classes such as risk exists in interest rate, risk exists in exchange rate, risk exists in equity, risk exists in commodity price and so on (Dowd, 2007).

How to manage the risk before estimating the value at risk is possible by various methods such as analysis of gap, analysis of duration, analysis of scenario, portfolio theory etc. Analysis of Gap will determine a particular time horizon, so that the price of our asset or liabilities can re-price with a better interest rate return, hence the choice of gap of time horizon is a very sensitive issue. The duration analysis is basically applied to the bond (the fixed income investment) to define the weighted average of the bond, where the weights are each class flow's present value which is related to all class flow's present value. Scenario analysis is a different approach where a scenario is set to examine whether we can stand to gain or loss under that scenario. Another approach is portfolio theory, where the investors have to choose the portfolio on the basis of their return on investment. Market risk is also called as systematic risk as it depends on various factors that have an impact on entire market such as recessions.

Market risk analysis is basically rising of uncertainty on the return of a portfolio's price (Alexander, 2009). To determine about the dispersion of the portfolios return from the target value, various portfolio's risk metrics is introduced such as volatility and correlation, but they are sufficient, when the risk factors return is having a multivariate normal distribution. Another interesting concept is about the premium of market risk, where the expected returns difference basically on a portfolio of market and risk-free rate is considered as a market risk premium. From a survey it is found that 54% of market risk premium used in 2016 is decreased, whereas 38% is increased (Fernandez, Ortiz & Acan, 2017). Alexander Brown *et al.* analyses over market risk standard formula, where the market risk capital charges is calculated under the standard formula (Braun, Schmeiser & Schreiber, 2017).

Liquidity Risk

Liquidity risk is the risk when a party paying attention to trading the financial asset, but no one is interested to trade for those financial assets. On the whole, an asset can be traded quickly to prevent the loss. Liquidity risk is essential for the party, who has held the asset since its ability is affected for trading. By dropping the asset's price from a particular price to zero means the asset is worthless. If a party will not find the interest of another party in trading the asset, then also in the emerging market, the liquid risk is found to higher. The risk of liquidity on financial market is basically happening due to the rise of uncertainty in liquidity. Liquidity risk can be occurred in an institution if its credit rating falls or some unexpected cash flows take place. When the other investor is facing lack of fund, then it is definitely difficult to sell the assets, here we can say the market and funding liquidity risks integrate each other. It also gives an attention to compound other type of risks. Whenever, a trading organization is having a position on a non-liquid asset, its talent to liquidate that position will be compounded its market risk. Liquidity risk can be compounded to credit risk, if a firm has offset cash flows with two different parties and the counter party owes it's a payment default, then to make payment, the farm has to raise cash from another party. It is difficult to isolate the liquidity risk as it is having a tendency to compound other risks. In addition to the market the liquid risk has to manage the credit and other types of risk. Liquidity risk can be judged in the form of scenario analysis, if the flow of cash of an organization is largely contingent. Scenario analysis is to estimate the probable value of a portfolio after a fixed period of time, expecting changes in the portfolio's security value such as interest rates change.

Liquidity risk can be measured by liquidity gap and elasticity. Liquidity gap is defined as a firm's net liquid asset that is the surplus value of the liquid assets of the firm over its volatile liabilities. In a statistical measure of liquidity risk, it does not give any indication to change the gap with increasing marginal cost of the firm. Another form of measure is elasticity, which is the change of net assets over funded liabilities that take place when the premium of liquidity on the marginal funding cost of the bank rises by a little amount, as the liquidity risk elasticity. Some researchers have analyzed over liquidity risk in various ways. Sperna Weiland, Laeven, and De Jong (2017) introduced a model, where, the author proves that between credit and liquidity risk there is a possibility of dynamic feedback loops. Xi Dong, Feng, and Sadka (2017) defined about the capability of fund managers to make value, which depends on the condition of market liquidity by introducing liquidity risk exposure for the skilled managers. On an analyzing over the data of 20 major organization, based on real equity and real currency prices a new evidence is reported such as at monthly frequency, relating to equity market a higher return is observed, which is related to foreign counterparts, by generally combining with home currency depreciation, but this relation which is correlated negatively can be broken down or even it can be reversed during the time of relatively higher aggregate economic uncertainty or volatility (Jung, 2017). Janek Gallitschke, Seifried, and Seifried (2017) presents a model based on interbank cash transactions and factorizing the interbank credit and liquidity risk. Eva Lütkebohmert analyzes about the increase in volatilities, where it founds when the volatility increases, the creditors turn into pessimistic in higher side, by which the fire sale rate decreases (Lütkebohmert, Oeltz & Xiao, 2017). In order to carry the creditors to overpower their findings, higher short-term rates are strongly needed. In this way illiquidity risk obtained basically from the result of volatilities increased and reinforces funding liquidity risk. Another aspect of liquidity risk is about the stock liquidity risk for pricing the asset and to determine the exposure of the firm to the crises arises for market liquidity is very important to understand that how the CEO equity incentives is having the influence over stock liquidity (Wruck & Wu, 2017).

Equity Price Risk

Equity risk refers to the price changes of equity shares associating with two different categories such as systematic risk and non-systematic risk. The equity risk measures the standard deviation of the price of security over a number of times. Next come description about systematic risk, which is caused by market factors affecting the entire industry by some events. Diversification is not possible in case of systematic risk. Unsystematic risk relates to a risk that arises to a specific company such as management changes, fraud etc. The equity share's price is highly affected by unsystematic risk. Suppose a new product is launched by a company, in terms of response to the new product, the market will have uncertainty due to the fluctuations arising in stock prices. Here, the risk is an unsystematic risk, which arises basically by the shareholders. The effective way to manage the equity price risk is the creation of a diversified portfolio, which includes securities having low or negative correlation among them. By this way one security's loss can be balanced with other security's gain.

Some of the research analysis is described here, where Maria Vassalou and Xing (2017). stated that the effects of size and book to market (BM) on a risk-based interpretation is related intimately to default risk. The effects of BM only exist for two quintiles base on highest default risk. The return difference of stock value (known as high book to market) and stock growth (known as low book to market) is 30% per year, and it moves down to 12.7% for the stock, considering the default risk of the second highest. BM does not have any effect on the market's remaining stock. In this category, the stock's value has the highest BMs among all the stocks with the smallest in size in the market. Kevin B. Hendricks and Vinod R. Singhal (2005) indeed stated in an article that after the announcement the value of equity risk 13.5% higher than the equity risk before the announcement in the year. In an article Wayne R Guay (1999) described about the control of risk related problems, where the author stated that the convexity and slope of the relationship between the performance of firm and the wealth of manager are managed by the equity holders. Arditti FD (1967) stated that equity risk variables are divided into two categories one associating directly with the probability distribution of returns such as the distribution of second and third moments and the existence of correlation between the single stock's return and stocks of other variables.

Foreign Exchange Risk

Foreign exchange risk is another category of market risk, which is associated with the fluctuations, occurred in the values of currency. It basically happens, in the case, when the financial transaction is dominated by a counter currency other than business's base currency. As US Dollar is the bench mark unit of currency, hence in most of the exchange rate basically they use US Dollar as the base currency and the counter currencies are the other currencies (Das, Mishra & Rout, 2017). The risk is handled by hedging the exposure in one currency to other currency, so that the fluctuations in currency exchange will not have any impact over transaction. To manage the foreign exchange risk, various types of instruments are available such as swapping of currency, money market hedges, future and forward contracts etc.

Bernard Dumas and Bruno Solnik (1995) investigated about the risk of exchange rate whether it is priced in international asset markets and the conditional approach which is allowed for the variation in time in the achievements for exchange rate risk. The existence of the premia belongs to foreign exchange risk is supported by the largest equity markets of the world. Another investigation has done by Ian Domowitz and Craig S. Hakkio (1985), where the author investigated about the risk premium's existence in the forex market based on the conditional variance of market forecast errors by following the ARCH process. Here the work is applied on currencies like France, Japan, United Kingdom, Switzerland and Germany. A multivariate generalized ARCH model is developed by Richard T. Baillie and Tim Bollersley (1990) for hypothesis test, that according to the suggestion of standard asset pricing theory literature, linear function belongs to the conditional variances and covariance. Robert J. Hodrick and Sanjay Srivastava (1984) examined the risk premium existence in the foreign exchange rate market. Based on the theoretical model of pricing of asset, the statistical model leads to severe cross-equation constraints, which can be rejected by statistical tests. Geert Bekaert and Robert J. Hodrick (1993) explored various sources belongs to measurement error and misspecification that might produce bias.

Commodity Price Risk

Another type of risk is commodity price risk, which basically link to the change of price of raw materials, which is needed in the business. The company's profit margin depends on the raw materials. For a potato chips company, the increase in the price of potato affects a lot on the cost of production of that company. This type of risk occurs in the commodity market. To manage the commodity price risk, the company holds the contracts for a long term supply. Including this type of managing commodity risk, other measures are also used to manage the risk such as increasing the costumer's price, looking for alternate commodity, or hedging with the different financial exposures.

Analyzing about the commodity risk, various researches explored by the researches are presented here. S. M. Lokare (2007) attempts to examine the efficiency and the performance of commodity derivatives in directing the price risk management.

The critical analysis about the performance reveal that the market yet to attain the minimum critical liquidity. Regarding the importance of commodity price in an article Söhnke M. Bartram (2015) stated that the commodity price is more fluctuating than the price of exchange rate and interest rate price. The author made a comprehensive analysis and from this critical analysis, it is found that though the commodity price is highly volatile in nature, but relating to risk, it is having less risk than financial risk. To manage the volatility in commodity market basically to manage the export of the earnings and import the payments, Theophilos Priovolos and Ronald C. Duncan (1991) showed some of financial instruments related to commodity price and Zvi Bodie and Victor Rosansky (1980) explored the commodity futures and found that the commodity futures proved to be very well in the inflation hedges.

Credit Risk

Credit risk occurs, basically due to the arise of fault on loans. Whenever the lenders lend money to the borrowers, there is a risk for the lenders that the borrowers may not repay the amount of loan. Credit risk is also known as default risk as the risk is associated with the borrowers who are not able for making payments. Suppose a company has borrowed a huge amount of money from the bank and unable to repay due to the losses occurred in the business, hence for the bank this poses credit default risk. The losses occurred in case of investors are loss of principal amount and the interest amount, decreased in the cash flow and increase in the collection of cost. By using the direct or indirect use of leverage, credit risk can be assumed by the investors. To manage the credit risk, various ways are available such as credit default swaps, which will provide protection against the credit loss.

A default risk is possible, when a counter party is unable to fulfill the commitment of the contracts (Bielecki & Rutkowski, 2013). This is called the default event occurs. If we will analyze the term "credit risk", then in the general term we can say the risk in linked with the credit, such as credit quality changes, variation occurred in the credit spreads and the default event. For pricing and hedging derivative securities Robert A. Jarrow and Stuart M. Turnbull (1995) proposed a new methodology involving the credit risk. Here, two types of credit risks are well thought-out by the author. The first one is, the asset underlying the derivative security may default and the second is the writer of the derivative security may default. Robert A. Jarrow proposed a markov model based on Jarrow and Turnbull (1995), with the process of bankruptcy subsequently following a discrete state space Markov chain in the credit ratings (Jarrow, Lando & Turnbull, 1997). Another model developed by Edward I. Altman and Anthony Saunders (1997), analyzing the risk return structure of portfolios by offering some promise, through exposing the debt instruments. The finance market and the credit risks are essentially related, they are inseparable.

Robert A. Jarrow and Stuart Turnbull (2000) explains about pricing the credit risky with respect to two main approaches one is structural approach and the other one is reduced form of approach.

Foreign Investment Risk

Foreign investment risk or currency exchange rate risk occurs due to the change in the price of one currency against the other currency. Here arises the concept of base currency and counter currency. The continuous fluctuation in foreign currency, where an investment is denoted in relation to one's home currency, which may add risk to the security's value. American investors convert any types of profits from foreign assets to the US Dollars. If the value of dollar is very strong then the foreign stock value or the bond which have purchased on a foreign exchange will turn down. This type of risk is augmented particularly, when one particular country's currency drops and the investments of everyone (those who have invested) are in foreign assets of that country. If the dollar value is weak, then the foreign assets of American investor will definitely rise. Arise of Risk in currency, basically happens for short term investment, as it does not have enough time to stabilize like the longer term for foreign investments. In case of Mutual fund also, whenever it invests in foreign securities, there is a chances for occurrence of a risk such as if a country is having foreign investment in a country then for mutual fund, to sell investment to that particular country is very difficult.

For taking decision the following things can be described: the first one is that the choice of anyone that the organization has made always depends on the social system, where the process has taken place, the second one is, the process takes a long time, though it is not composed of each one's decision, third one is, the decisions are made with no certainty, fourth one is, the organization should have a goal, and the last one is, so many constraints are available on the decision maker's freedom (Buckley & Ghauri, 2015). Elizabeth Asiedu (2002) explores about the factors that effect on foreign direct investment in the developing countries, whether it is affecting in sub-Saharan Africa in a different way. The result shows: (a) a higher investment returns and with a better infrastructure, is having a positive impact on foreign direct investment to non-sub-Saharan Africa countries; significant impact over foreign direct investment to non-sub-Saharan Africa is not possible, (b) Openness for trading, promotes foreign direct investment to sub-Saharan Africa and non- sub-Saharan Africa countries. However, the marginal benefits are very less for sub-Saharan Africa, from the increased openness. Llewellyn D. Howell and Brad Chaddick (1994) explored over the effect of political risk methods, by studying a comparison on the projections of 1986 of the economists, political risk services and BERI in opposition to losses acquire in the period of 1987–1992. The discussion focused on the losses of foreign investors and the source of losses. Considering the foreign direct investment Llewellyn D. Howell stated that, the direct investment in foreign occurs basically in the industries, which is characterized by lending and borrowing countries (Cruz, 2002).

Operational Risk

Operational risk arises by the failure of internal process like system failure, people failure, failure from the external events etc. Operational risk does not inherent from financial risk, systematic risk or market risk, rather this risk is remaining after the determination of financial risk as well as systematic risk. In a broad way operational risk is very close to good management and quality management. The satisfaction of the client, its reputation and the value of shareholder, all are affected by the operational risk. Unlike credit risk and market risk, operational risk is not revenue driven. Even operational risk cannot be interrupted temporarily as long as there exist imperfection in system, people and processes. Operational risk can be manageable within a risk tolerance zone and the tolerance is determined by balancing the improvement cost against the expected benefits. It can be summarized as the risk occurs due to human error, the risk that takes place due to the failing in business operations. In that scenario it varies from industry to industry. The industry having lower human interactions are likely to have less operational risk. Operational risk concentrates on, how the things are carried out within an organization, without considering the production within an industry. Decisions regarding organization's priority and organizational function are also associated with operational risk.

Relating to the definition of Basel, Marcelo G Cruz stated operational risk as the risk of losses, which arises from internal controls, external events, failure of systems as well as people, as most of the financial institutions are not structured properly (Cruz, 2002). The consequences that arise from the operating risk relates to the postponement of payment for good as well as services (Muchalski, 2008). Considering this consequences Grzegorz Michalski presented a method, which uses the theory of portfolio management and determined the accounts receivable level in a firm. Valérie Chavez-Demoulin, Paul Embrechts, and Johanna Nešlehová (2006) presented some probability and statistics techniques to prove its usefulness in any type of quantitative modeling environment. In the other area of quantitative risk management, the above technique can be also be applied.

Exploring the statistical behavior of the financial data Marco Moscadelli (2004) increased the understanding level of operational risk within the financial system and derived the bottom-up capital charge figures. Another author Johanna Nešlehová, Paul Embrechts, and Valérie Chavez-Demoulin (2006) explored on statistical analysis of operational risk data and raised some of issues relating to the correlation or diversification effects, the basic use of extreme value theory and the overall quantitative risk management consequences of extremely heavy tailed data. Operational risk is something that should be calculated actively to meet the objectives of shareholders, customers, management and stake holders. The objective includes the survival of the company or the secured future of the company, downgrades of the company by rating agencies should be avoided and the remaining solvent should come for many years (Panjer, 2006). In financial service industry operational risk performs a major part.

Model Risk

Model risk occurs due the inaccurate model, which is insufficient to give the result of market risk. Basically, the model is insufficient to take the accurate decisions value transaction. Other than the financial securities, model risk is prevalent in many other fields such as assigning the consumer credit scores, calculating the probability of an air flight passenger, that he is a terrorist. Model risk is well thought-out as a subset of operational risk, as the firm that creates as well as uses model is affected by the model risk. The limitations of the model occurred, when the investors or the traders does not understand completely about the assumptions and limitations of the model. In the past decade, the financial model is very prevalent with a new type of financial securities.

T. Clifton Green and Stephen Figlewski (1999) indicated about the chances of fairly large risk, by occurring the imperfect models and inaccurate results obtained in the volatility forecasts. In this article the occurrence of damage to a maximum extent, due to the model risk can be limited by using the pricing options with the use of higher volatility than estimating the best from historical data. In an article Grahame R. Dowling and Richard Staelin (1994) focused on the importance of determining the risk handling activity by perceiving the benefits of the various types of activity for handling the risk and inability occurred by the consumers to absorb the monetary loss.

Financial Risk

Financial risk occurs due to the possibility of shareholders to lose money at the time of investing in a company, having debt and also when the cash flow of the company proves to be insufficient to meet the financial obligations. The company, who uses debt financing, if it becomes insolvent, the creditor has to repaid before the shareholders. In general term, financial risk is related to finance industry, which includes risk that occurs due to financial transactions like company loans, and its disclosure to loan default. The financial risk includes liquidity risk, foreign investment risk, currency risk, credit risk etc. The above types of risks are already discussed in respective sections.

Some of the researches done by the researches, where Lyn C Thomas (2000) discussed about the necessity of incorporating the economic conditions in to scoring systems and the way of changing of the system to estimate the probability of consumer's default to estimate the consumer's profit will bring to the lending organization. Stephen C. Gabriel and Chet B. Baker (1980) stated in an article that the financial risk add net cash flows variability of the equity's owners, which can be obtained as a result getting from the fixed financial obligations and linked with debt financing and also cash leasing.

APPROACHES TO SOLVE THE ISSUES ARISING IN THE FORM OF RISK IN THE FINANCIAL MARKET

Some approaches have been described for getting the solution of the risk arising in financial market, which is analyzed in the previous section, are given below.

Market risk is the uncertainty about the financial market. By creating a real time risk management frame work for analyzing and quantifying the market risk, developing a strategy for managing the market risk and developing appropriate process that links with commodity policy, energy price police, for supporting ongoing management of market risk, the market risk can be mitigated.

To successfully manage liquidity risk, it needs the establishment of a risk management framework, which should be robust and make certain that having sufficient liquidity. It includes high quality liquid assets with the aim of endure stress events, which includes both secured as well as unsecured funding sources, involving loss or injury for both secured and unsecured.

Equity price risk cannot be known specifically, as no one knows about the future performance of a particular stock or stock market. It can be estimated ridiculously by depending on the time frame and also through calculating using methods. It can also be roughly calculated as a backward-looking quantity through stock market and the performance of government bond over a defined time period, like 1980 to present.

Though foreign exchange risk cannot be avoided as it is a risk of foreign investing but by using hedging techniques it can be considerably mitigated. If we want to totally eradicate the forex risk, then the choice of investing in overseas should be avoided. But from the portfolio of diversification, this is not the alternative, as in several studies it is described that foreign investing improves the portfolio return at the same time as reducing the risk.

To manage the Commodity price, companies has to manage increasing commodity prices through the physical hedges for buying enough time for the modification of their pricing models and the pass-through arrangement can be arranged for the customer.

Credit risk is occurring due to failure of meeting the obligations in keeping the agreed term, which can be solved by fulfilling regulatory necessities, and at the same with a same stream line process with a speed lending decision. Its operational efficiency and increase in cost needs automation. The data should be available around the clock, so that reporting of credit risk cannot be hindered by problematic delays.

There are four ways which have described here, to overcome from foreign investment risk. First one is the hedging foreign assets in the portfolio; if the currency of investment is falling then no money will be loosed, though no currency will be gained also. The second one is when one currency is depreciated, then that currency should be sold, and another currency should be purchased. In third, it is described to purchase the currency of a country with a high rate of interest, as just by holding the country's currency with high investment return in a bank account or by possessing the government bonds, 4 percent of money for annually can be hold. The forth one is not to buy over valued currencies, as if the country has a high inflation rate relative to other country, then country with high inflation rate will lose its competitiveness after a while.

Operational risk can be mitigated by making a robust frame work to identify and manage both operational as well as financial reporting risk. By which the shareholder value can be protected, and the business performance can be maximized.

As a defective model is the cause of one leading financial institution to undergo losses, when a coding error contorted the flow of information from the risk model towards the process of portfolio optimization. By designing the standard model with advanced analytics techniques, model risk can be reduced, as incorrect models can cause a great harm.

Financial risk is always associated with field of investment. The investment can lose money, or it can grow up, as it is always a chance. Hence to reduce the financial risk, the management of investment portfolio should be learned properly.

CONCLUSION

Various types of risk arise in the financial market, which creates the possibility of losing money in various way such as loss of money by investing through share market, investing in forex market, the clash flows of a company proving insufficient to meet its financial obligations etc. In this study, a variety of risk occurs in financial market is analyzed and an effective solution for this risk is well thought out. In future these risks arise in financial market can be solved using soft computing techniques.

REFERENCES

Alexander, C. (2009). *Market risk analysis, value at risk models* (Vol. 4). John Wiley & Sons.

Altman, E. I., & Saunders, A. (1997). Credit risk measurement: Developments over the last 20 years. *Journal of Banking & Finance, 21*(11), 1721–1742. doi:10.1016/S0378-4266(97)00036-8

Arditti, F. D. (1967). Risk and the required return on equity. *The Journal of Finance, 22*(1), 19–36. doi:10.1111/j.1540-6261.1967.tb01651.x

Asiedu, E. (2002). On the determinants of foreign direct investment to developing countries: Is Africa different? *World Development, 30*(1), 107–119. doi:10.1016/S0305-750X(01)00100-0

Bartram, S.M. (2015). *The Impact of Commodity Price Risk on Firm Value--An Empirical Analysis of Corporate Commodity Price Exposures.* Academic Press.

Bekaert, G., & Hodrick, R. J. (1993). On biases in the measurement of foreign exchange risk premiums. *Journal of International Money and Finance, 12*(2), 115–138. doi:10.1016/0261-5606(93)90019-8

Bielecki, T. R., & Rutkowski, M. (2013). *Credit risk: modeling, valuation and hedging.* Springer Science & Business Media.

Bodie, Z., & Rosansky, V. I. (1980). Risk and return in commodity futures. *Financial Analysts Journal, 36*(3), 27–39. doi:10.2469/faj.v36.n3.27

Braille, R. T., & Bollersley, T. (1990). A multivariate generalized ARCH approach to modeling risk premia in forward foreign exchange rate markets. *Journal of International Money and Finance, 9*(3), 309–324. doi:10.1016/0261-5606(90)90012-O

Braun, A., Schmeiser, H., & Schreiber, F. (2017). Portfolio optimization under solvency II: Implicit constraints imposed by the market risk standard formula. *The Journal of Risk and Insurance, 84*(1), 177–207. doi:10.1111/jori.12077

Buckley, P. J., & Ghauri, P. N. (2015). *International business strategy: theory and practice.* Routledge. doi:10.4324/9781315848365

Chavez-Demoulin, V., Embrechts, P., & Neslehova, J. (2006). Quantitative models for operational risk: Extremes, dependence and aggregation. *Journal of Banking & Finance, 30*(10), 2635–2658.

Clifton Green, T., & Figlewski, S. (1999). Market risk and model risk for a financial institution writing options. *The Journal of Finance, 54*(4), 1465–1499.

Cruz, M. G. (2002). *Modeling, measuring and hedging operational risk*. Chichester, UK: John Wiley & Sons.

Das, S.R., Mishra, D., & Rout, M. (2017). A hybridized ELM-Jaya forecasting model for currency exchange prediction. *Journal of King Saud University-Computer and Information Sciences.*

Domowitz, I., & Hakkio, C. S. (1985). Conditional variance and the risk premium in the foreign exchange market. *Journal of International Economics*, *19*(1-2), 47–66. doi:10.1016/0022-1996(85)90018-2

Dong, X., Feng, S., & Sadka, R. (2017). Liquidity risk and mutual fund performance. *Management Science*, mnsc.2017.2851. doi:10.1287/mnsc.2017.2851

Dowd, K. (2007). *Measuring market risk*. John Wiley & Sons.

Dowling, G. R., & Staelin, R. (1994). A model of perceived risk and intended risk-handling activity. *The Journal of Consumer Research*, *21*(1), 119–134. doi:10.1086/209386

Dumas, B., & Solnik, B. (1995). The world price of foreign exchange risk. *The Journal of Finance*, *50*(2), 445–479. doi:10.1111/j.1540-6261.1995.tb04791.x

Fernandez, P., Ortiz, A., & Acan, I. F. (2017). Market Risk Premium Used in 71 Countries in 2016: A Survey with 6,932 Answers. *Journal of International Business Research and Marketing*, *2*(6), 23–31. doi:10.18775/jibrm.1849-8558.2015.26.3003

Gabriel, S. C., & Baker, C. B. (1980). Concepts of business and financial risk. *American Journal of Agricultural Economics*, *62*(3), 560–564. doi:10.2307/1240215

Gallitschke, J., Seifried, S., & Seifried, F. T. (2017). Interbank interest rates: Funding liquidity risk and XIBOR basis spreads. *Journal of Banking & Finance*, *78*, 142–152. doi:10.1016/j.jbankfin.2017.01.002

Guay, W. R. (1999). The sensitivity of CEO wealth to equity risk: An analysis of the magnitude and determinants. *Journal of Financial Economics*, *53*(1), 43–71. doi:10.1016/S0304-405X(99)00016-1

Hendricks, K. B., & Singhal, V. R. (2005). An empirical analysis of the effect of supply chain disruptions on long-run stock price performance and equity risk of the firm. *Production and Operations Management*, *14*(1), 35–52. doi:10.1111/j.1937-5956.2005.tb00008.x

Hodrick, R. J., & Srivastava, S. (1984). An investigation of risk and return in forward foreign exchange. *Journal of International Money and Finance, 3*(1), 5–29. doi:10.1016/0261-5606(84)90027-5

Howell, L. D., & Chaddick, B. (1994). Models of political risk for foreign investment and trade: An assessment of three approaches. *The Columbia Journal of World Business, 29*(3), 70–91. doi:10.1016/0022-5428(94)90048-5

Jarrow, R. A., Lando, D., & Turnbull, S. M. (1997). A Markov model for the term structure of credit risk spreads." -. *Review of Financial Studies, 10*(2), 481–523. doi:10.1093/rfs/10.2.481

Jarrow, R. A., & Turnbull, S. M. (1995). Pricing derivatives on financial securities subject to credit risk. *The Journal of Finance, 50*(1), 53–85. doi:10.1111/j.1540-6261.1995.tb05167.x

Jarrow, R. A., & Turnbull, S. M. (2000). The intersection of market and credit risk. *Journal of Banking & Finance, 24*(1), 271–299. doi:10.1016/S0378-4266(99)00060-6

Jung, K. M. (2017). liquidity risk and time-varying correlation between equity and currency returns. *Economic Inquiry, 55*(2), 898–919. doi:10.1111/ecin.12418

Lokare, S. M. (2007). Commodity derivatives and price risk management: An empirical anecdote from India. *Reserve Bank of India Occasional Papers, 28*(2), 27-76.

Lütkebohmert, E., Oeltz, D., & Xiao, Y. (2017). Endogenous credit spreads and optimal debt financing structure in the presence of liquidity risk. *European Financial Management, 23*(1), 55–86. doi:10.1111/eufm.12089

Michalski, G. (2008). *Operational Risk in Current Assets Investment Decisions: Portfolio Management Approach in Accounts Receivable* [Agro Econ-Czech: Operační Risk v Rozhodování o Běžných Aktivech: Management Portfolia Pohledávek]. Academic Press.

Moscadelli, M. (2004). *The modelling of operational risk: experience with the analysis of the data collected by the Basel Committee*, Academic Press.

Neslehova, J., Embrechts, P., & Chavez-Demoulin, V. (2006). Infinite mean models and the LDA for operational risk. *Journal of Operational Risk, 1*(1), 3–25. doi:10.21314/JOP.2006.001

Panjer, H. H. (2006). *Operational risk: modeling analytics* (Vol. 620). John Wiley & Sons. doi:10.1002/0470051310

Priovolos, T., & Duncan, R. C. (1991). *Commodity risk management and finance*. The World Bank.

Sperna Weiland, R. C., Laeven, R. J. A., & De Jong, F. (2017). Feedback between credit and liquidity risk in the us corporate bond market. *30th Australasian Finance and Banking Conference 2017.*

Thomas, L. C. (2000). A survey of credit and behavioural scoring: Forecasting financial risk of lending to consumers. *International Journal of Forecasting, 16*(2), 149–172. doi:10.1016/S0169-2070(00)00034-0

Vassalou, M., & Xing, Y. (2004). Default risk in equity returns. *The Journal of Finance, 59*(2), 831–868. doi:10.1111/j.1540-6261.2004.00650.x

Wruck, K.H., & Wu, Y.L. (2017). *Equity incentives, disclosure quality, and stock liquidity risk*. Academic Press.

Chapter 9

Clustering Techniques:
A Review on Some Clustering Algorithms

Harendra Kumar
Gurukula Kangri Vishwavidyalaya, India

ABSTRACT

Clustering is a process of grouping a set of data points in such a way that data points in the same group (called cluster) are more similar to each other than to data points lying in other groups (clusters). Clustering is a main task of exploratory data mining, and it has been widely used in many areas such as pattern recognition, image analysis, machine learning, bioinformatics, information retrieval, and so on. Clusters are always identified by similarity measures. These similarity measures include intensity, distance, and connectivity. Based on the applications of the data, different similarity measures may be chosen. The purpose of this chapter is to produce an overview of much (certainly not all) of clustering algorithms. The chapter covers valuable surveys, the types of clusters, and methods used for constructing the clusters.

INTRODUCTION

Clustering is the task of dividing the set of data points (populations) into a number of groups (clusters) such that data points in the same groups are more similar to other in the same group than those data points in other groups. Or clustering is a process of organizing data points into groups whose members are similar in some way. A cluster is therefore a collection of objects which are "similar" between them and are "dissimilar" to the objects belonging to other clusters. Clustering can be considered the most important unsupervised learning problem as it finds a structure in a collection of unlabeled data. Such grouping is pervasive in the way human

DOI: 10.4018/978-1-5225-5793-7.ch009

process information, and one of the motivations for using clustering algorithms is to provide automated tools to help in constructing categories or taxonomies. These formed clusters are used as the basis for the data analysis (or data processing techniques). The formed clusters include dense areas of the data space, groups with small distances, cluster data members, intervals or particular statistical parameter. Therefore, a data clustering can be formulated as a multi-objective optimization problem. The selection of appropriate clustering algorithm and parameter settings depend on the types of data set and intended used for the results.

Almost all clustering problem is NP hard for almost all clustering objective functions. Distance-based and density-based are two of many categories of clustering algorithms. The distance-based clusters are formed by adding points that minimize intra-cluster distances and maximize inter-cluster distances. The radius and diameter of clusters can be used as intra-cluster characteristics. Density based clustering help searches for signals of unknown shape. In a density-based clustering, a cluster is connected dense component which can grow in any direction to increase the density. Density based algorithm looks for neighbours of those data points that have at least a given number of neighbouring points within a given distance on the plane and forms clusters of data-points that can be related through their common neighbours.

Clustering algorithms work very differently so it is very difficult to conclude which algorithm is the best without examining the formed data clusters. Besides choosing the right clustering algorithm, choosing the right features also plays a critical role in clustering. Moreover, there are no universally accepted and effective criteria for selecting the clustering schemes and valid features. Although, validation criteria can provide some insights on the quality of clustering solutions but even how to choose the appropriate criterion is still a problem requiring more research to do. Han and Kamber (2001) have given a very good introduction to contemporary data mining clustering techniques in their text book. Genther et al. (1994) presented a modified fuzzy clustering algorithm for parametric defuzzification in fuzzy rule base systems. Some recent defuzzification methods have been discussed by Kumar (2017). The general discussion about hierarchical clustering is available in most of the clustering books. Zahn (1971) has discuss about divisive hierarchical clustering techniques that uses the minimum spanning tree of graph. Leung et al. (2000) have derived an interesting hierarchical clustering technique which is based on scale-space theory by using a blurring process, in which each datum is regarded as a light point in an image and a cluster is represented as a blob. Morzy et al. (1999) introduce a hierarchical algorithm that uses sequential patterns as the basic element found in the database to efficiently generate data clusters and defined a co-occurrence measure, as the standard of fusion of smaller clusters. Selim and Ismail (1984) gave rigorous proof of the finite convergence of the *K*-means type algorithms.

Purposes of Clustering

The quality of the data clustering result depends on its implementation and similarity measure used by the method. Also, the quality of a good clustering technique is measured by its ability to find some or all of the hidden patterns. Some purposes of a good clustering are:

1. To analyze the structure of the data;
2. To assist in classification designing
3. To relate different aspects of the data to each other.
4. To shape and keep knowledge.

Types of Measures

Since clustering is the grouping of similar data points (objects), there are two main types of measures used to estimate whether two objects are similar or dissimilar are distance measures and similarity measures. Distance measures to determine the similarity or dissimilarity between any pair of objects are used in several clustering methods. Given a set of input data points $X = \left\{ x_1, x_2, x_3 \ldots x_j, \ldots x_n \right\}$, where $x_j = \left(x_{j1}, x_{j2}, x_{j3} \ldots x_{jm} \right)^T \in R^m$ each measure x_{ji} is said to be a feature (attribute, dimension, or variable). The distance between two data points x_i and x_j is generally denoted by $d\left(x_i, x_j \right)$. The distance measure is said to be a metric distance measure if it also satisfies the following properties:

1. $d\left(x_i, x_j \right) \geq 0$, $\forall x_i, x_j \in X$
2. $d\left(x_i, x_j \right) \leq d\left(x_i, x_k \right) + d\left(x_k, x_j \right)$, $\forall x_i, x_j, x_k \in X$
3. $d\left(x_i, x_j \right) = d\left(x_j, x_i \right)$, $\forall x_i, x_j \in X$

Given two m-dimensional instances, $x_i = \left(x_{i1}, x_{i2}, \ldots x_{im} \right)$ and $x_j = \left(x_{j1}, x_{j2}, \ldots x_{jm} \right)$. Some different distance measures for clustering are given in Table 1.

Types of Clustering

Cluster algorithms can be categorized based on how the underlying models operate. As there are possibly 150 plus published algorithms for data clustering and not all of them provide models for their clusters. Thus clustering techniques not easily can

Table 1. Different distance measures for clustering

S.No.	Measure	Form	Comment
1.	Minkowski distance	$d_{ij} = \left(\sum_{l=1}^{n} \left| x_{il} - x_{jl} \right|^{p} \right)^{1/p}$	Standardization is necessary, if scales differ.
2.	Euclidean distance	$d_{ij} = \left(\sum_{l=1}^{n} \left| x_{il} - x_{jl} \right|^{2} \right)^{1/2}$	Special case of Minkowski distance for n=2.
3.	City-block distance	$d_{ij} = \sum_{l=}^{n} \left| x_{il} - x_{jl} \right|$	Special case of Minkowski distance for n=1.
4.	Sup distance	$d_{ij} = \max_{1 \le i \le n} \left| x_{il} - x_{jl} \right|$	Special case of Minkowski distance for $n \to \infty$.
5.	Chebyshev distance	$d(x, v) = \max_{i=1,2,\dots n} \left| x_i - v_i \right|$	Where $d(x, v)$, represent the distance between data vector x and centroid v.
6.	Chi-square distance	$d(x, v) = \sqrt{ \sum_{i=1}^{n} \frac{(x_i - v_i)^2}{x_i + v_i} }$	Where $d(x, v)$, represent the distance between data vector x and centroid
7.	σ -distance	$d(x, v) = \dfrac{x - v}{\sigma_i}$	Where σ_i is the weighed mean distance in cluster i.
8	Mahalanobis distance	$d_{ij} = \left(x_i - x_j \right)^{T} S^{-1} \left(x_i - x_j \right)$	Where S is within group covariance matrix
9	Cosine similarity	$S_{ij} = \cos \alpha = \dfrac{x_i^T x_J}{x_i x_j}$	The most commonly used in document clustering
10	City Block (Manhattan) distance	$d_{ij} = \sum_{j=1}^{m} \left(x_j - y_j \right)$	Where x_j and y_j are two points of m dimension

be categorized. But on the bases of cluster structure, the following two categories provide a baseline overview of clustering.

1. **Hard Clustering:** In hard clustering each data point is assigned to one and only one cluster. It attempts to seek a k-partition of X, $C = \{C_1, C_2, \ldots C_c\}(c \leq n)$ such that:

   a. $C_i \neq \varnothing, i = 1, 2, \ldots c$;

   b. $\bigcup_{i=1}^{c} C_i = X$;

   c. $C_i \cap C_j = \varnothing, i, j = 1, 2, \ldots c$ and $i \neq j$. Hard clustering also can be divided into two types hierarchical and partitional clustering.

      i. **Hierarchical Clustering:** Hierarchical clustering attempts to construct a tree-like nested structure partition of X, $H = \{H_1, H_2, \ldots H_Q\}(Q \leq n)$ such that $C_i \in H_p, C_j \in H_q$ and $p > q$ imply $C_i \in C_j$ or $C_i \cap C_j = \varnothing$ for all $i, j \neq i, p, q = 1, 2, 3, \ldots Q$.

      ii. **Partitional Clustering:** Partitional clustering decomposes a set of data points into a set of disjoint clusters. In partitional clustering each pattern only belongs to one cluster. However, a pattern may also be allowed to belong to all clusters with a degree of membership $\mu_{i,j} \in [0,1]$ which represents the membership coefficient of the j-th object in the i-th cluster and satisfies the following two constraints:

$$\sum_{i=1}^{c}\mu_{i,j} = 1, \text{ for all j and } \sum_{j=1}^{n}\mu_{i,j} < n, \text{ for all i}$$

   k-means is one of the highly used partitional clustering algorithms. The algorithm clusters the *n* data points into *k* groups by defining *k* centroids, one for each cluster. For this, *k* data points are selected at random from sample points as initial centroids. In the next step, each point belonging to the given data set is assigned to its nearest centroid. Euclidean distance is generally considered to calculate the distance between data points and the centroids.

2. **Model Based Clustering:** The main disadvantage of partitional and hierarchical clustering algorithms is that they are largely heuristic and not based on formal models. In model based clustering, the set of data points is modelled by using a standard statistical model to work with different distribution. In model-based clustering method, maximum likelihood estimation is used to

find the parameter inside the probability model. Banfield and Raftery (1993) is the classic reference for model based clustering. Fraley and Raftery (1999) discussed a comprehensive mixture- model based clustering scheme which was implementedas a software package, known as MCLUST. Xu *and* Wunsch (2005) have discussed various model based clustering models in their survey work. Melnykov and Maitra (2010) have provided a detailed review into mixture models and model-based clustering.

Another one possible classification of clustering can be according to whether the data sets are fuzzy (soft) or crisp (hard).

- **Hard Clustering:** In a hard clustering, the set of data points is partitioned into a specified number of mutually exclusive subsets. Hard clustering has the following advantages:
  - ○ Applicable to fairly large data sets.
  - ○ Relatively scalable and simple
- **Fuzzy Clustering:** In a fuzzy clustering, a data point can belong to several clusters simultaneously with different degrees of membership. In many situations, fuzzy clustering is more natural in comparison of hard clustering. Data points on the boundaries between several clusters are not forced to fully belong to one of the cluster, but rather are assigned membership degrees between 0 and 1 indicating their partial membership in a cluster. Fuzzy clustering has the following advantages:
  - ○ A data point can belong to multiple clusters
  - ○ Different clusters get formed for different idea
  - ○ Easy of computation and simplicity
  - ○ It can handle uncertainty at various stages

Clustering Algorithm

The choice of a suitable clustering algorithm and the choice of a suitable measure for the evaluation depend on the clustering task and objects. There is no clustering algorithm available so far that can be universally used to solve all problems. In this section, we discuss some popular and mostly used clustering algorithms.

k-Means Clustering [KMC] Algorithm

k-means clustering algorithm is one of the simplest and widely used unsupervised learning algorithms for solving the clustering problem. The term "k-means" was first used by James MacQueen in 1967 by using the idea given by Hugo Steinhaus in 1957. The aim of the KMC is to partition n data points into k clusters, $\{C_1, C_2, \ldots C_k\}$ represented by their centers or means. The center of each cluster is calculated as the mean of all the instances belonging to that cluster. For k-means we use the notion of a centroid, which is the mean or median point of a group of data points.

Let $X = \{x_1, x_2, \ldots x_n\}$ be the set of data points and $v = \{c_1, c_2, \ldots c_k\}$ be the set of cluster centers. KMC algorithm tries to make partitioning (or clusters) of n data points into k disjoint subsets C_j of data points. The KMC algorithm aims to minimize the sum-of-squares criterion;

$$J = \sum_{J=1}^{k} \sum_{x \in C_j} \left| x_n - c_j \right|^2,$$

where x_n is a vector representing the the nth data point and c_j is the geometric centroid of the data points belonging in cluster C_j. The clustering is done by minimizing the sum of squares of distances between data points and the corresponding cluster centroid. The steps of the KMC algorithm are described as follows:

Step 1: Begin with a decision on the value of k = number of clusters.
Step 2: Make an initial partition that make partition of data points into k clusters.
Step 3: Compute the distance of each data point from each of the cluster centroids.
Step 4: Assign the data point in a cluster whose distance from the cluster center is minimum of all the cluster centers.
Step 5: Update the new cluster center for each cluster.
Step 6: Repeat steps 3-6 until convergence is achieved.

Although KMC algorithm has the great advantage of being easy to implement, it also has two big drawbacks. First, it can be really slow, as it calculates the distance between each point to each cluster in each step, which can be really expensive for the large dataset. Second, KMC algorithm is quite sensitive to the initial clusters. However, in recent years, this problem has been addressed with some degree of success. It does not work well with clusters of different size of different density. The KMC algorithm and many of its variations are described by Anderberg (1973) and Jains and Dubes (1988) in detail in their books.

k- Medoids Clustering [KMEC] Algorithm

In some applications, we want each center to be one of the points itself. The *k*-medoids clustering algorithm is related to the *k*-means algorithm. Both the KMC and KMEC algorithms partitions set of data points into *k* number of mutually exclusive clusters. These both the clustering techniques assign each data point to a cluster by minimizing the distance from the data point to the mean or median location of its assigned cluster, respectively. k-medoids is similar to k-means clustering algorithm, except when fitting the centres $c_1, c_2, \ldots c_k$. Instead of using the mean point as the center of a cluster, *k*-medoids uses an actual point in the cluster to represent it. The basic idea of KMEC algorithm is to first compute the k representative data points which are called as medoids. After finding the set of medoids, each data point of the data set is assigned to the nearest medoid. The steps of the KMC algorithm are described as follows:

Step 1: Initially select k random points as the medoids from the given n data points of the data set.

Step 2: Assign each remaining data point in a cluster having nearest medoid.

Step 3: Randomly select a non-medoid data point and compute the total cost of swapping old medoid data point with the currently selected non-medoid data point.

Step 4: If the total cost of swapping is less than zero, then perform the swap operation to generate the new set of k-medoids.

Step 5: Repeat the steps 2-4 until there is no change of the medoids.

Kaufman and Rousseeuw (1987) discussed a novel approach of k-medoids clustering. Park and Jun (2009) proposed a new algorithm for KMEC which runs like the KMC algorithm. Jin and Han (2010) have shown the difference between means and medoids clustering algorithms in a 2-D with examples.

Fuzzy c-Means Clustering [FCMC] Algorithm

Fuzzy clustering (soft clustering) is a form of clustering in which each data point can belong to more than one clusters simultaneously with different degrees of membership. Fuzzy c-means is one of the most widely used and a powerful unsupervised method to analysis the data and to construct the models. Fuzzy c-means (FCM) clustering was developed by J.C. Dunn in 1973, and later on in 1981 it was improved by J.C. Bezdek. It is useful when the required numbers of clusters are pre-determined. Let us suppose that m-dimensional n data points represented by x_i

(i = 1, 2,...,n), are to be clustered in c cluster. The algorithm returns a list of $c\{\backslash displaystyle c\}$ cluster centres $C = \{c_1, c_2, ..., c_c\}$ and a partition matrix $U = [u_{i,j}] \in [0,1]$, $i = 1, 2, ..., n, j = 1, 2, ..., c$, where $u_{i,j}$ tells the degree to which element x_i belongs to cluster C_j. Here $u_{i,j} \in [0,1]$ and $\sum_{j=1}^{c} u_{i,j} = 1, \forall i$. The FCMC algorithm aims to minimize the flowing objective function:

$$J_m = \sum_{i=1}^{n} \sum_{j=i}^{c} u_{i,j}{}^m x_i - c_j{}^2,$$

where m is fuzzy partition matrix exponent for controlling the degree of fuzzy overlap.

The steps of the FCMC for partitions the m-dimensional data set $X = \{x_1, x_2, ..., x_n\}$ into c fuzzy clusters are follows:

Step 1: Assume the number of clusters, c to be made, where $2 \leq c \leq n$

Step 2: Randomly, initialize the cluster membership values $u_{i,j}$.

Step 3: Calculate the cluster centres:

$$c_j = \frac{\sum_{i=1}^{n} u_{i,j}{}^m x_i}{\sum_{i=1}^{n} u_{i,j}{}^m}$$

Step 4: Update cluster membership values $u_{i,j}$ according to the following:

$$u_{i,j} = \frac{1}{\sum_{k=1}^{n} \left(\dfrac{x_i - c_j}{x_i - c_k} \right)^{\frac{2}{m-1}}}$$

Step 5: Calculate the objective function, J_m.

Step 6: If J_m improves by less than a specified minimum threshold (or a specified maximum number of iterations) then go to step 7, else go to step 3.

Step 7: Stop.

FCMC algorithm works by assigning membership to each data point on the basis of distance between the data point and cluster center. It works with the concept that, more the data is near to the cluster center more is its membership towards the particular cluster center. Clearly, the summation of membership values of each data point in a cluster must be equal to one. Here, tolerance of the required clustering is measured by fuzziness coefficient m, where $1 < m < \infty$. The value m determines how much the clusters can overlap with one another. The higher the value of m indicates the larger overlapping between clusters.

Possibilistic c-Means Clustering [PCMC] Algorithm

Krishnapuram and Keller (1993) proposed a possibilistic clustering approach named possibilistic c-means clustering [PCMC] to overcome the limitations of FCMC algorithm. In PCMC algorithm, $u_{i,j} \in [0,1]$ and memberships of a pattern to all the c clusters are not constraint to sum up to one. In other words, in PCMC, the following constraint:

$$\sum_{j=1}^{c} u_{i,j} = 1, \forall i$$

is relaxed which is known as probabilistic constraint,.

Let us suppose that M-dimensional n data points represented by x_i (i = 1, 2,...,n), are to be clustered into c number of clusters (1 < c < n). The algorithm returns a list of $c\{\backslash displaystyle\,c\}$ cluster with centres $C = \{c_1, c_2,...,c_c\}$. The objective function for PCMC algorithm is given by:

$$J_m = \sum_{j=1}^{c}\sum_{i=1}^{n} u_{i,j}^m d_{i,j}^2 + \sum_{j=1}^{c}\eta_j \sum_{i=1}^{n}\left(1 - u_{i,j}\right)^m$$

where $\eta_j > 0 \left(0 \le j \le c\right)$.

By using an approximate optimization of J_m, the updated formulas for membership degree $u_{i,j}$ and centre c_i are:

$$u_{i,j} = \frac{1}{1 + \left(\dfrac{d_{i,j}^2}{\eta_j}\right)^{\frac{1}{m-1}}}, \forall i, j$$

$$c_j = \frac{\sum_{i=1}^{n} u_{i,j}^m x_i}{\sum_{i=1}^{n} u_{i,j}^m}, \forall j$$

The membership of x_j to cluster i depends only on $d_{i,j}$ to this cluster. A small distance corresponds to a high degree of membership. Larger distances result in low membership degrees.

Estimation of η_j : The value of parameter, η_j is estimated by using the following equation:

$$\eta_j = K \frac{\sum_{i=1}^{n} u_{i,j}^m d_{i,j}^2}{\sum_{i=1}^{n} u_{i,j}^m}, K > O$$

Intuitively, η_j is an estimate of the spread of the i-th cluster and the value of K can be set to have a better control on it. In most of the cases the value of K is taken one. The PCMC algorithm is more effective in the presence of noise and it efficiently finds valid clusters.

Fuzzy Possibilistic c-Means Clustering [FPCMC] Algorithm

To overcome difficulties of the PCMC algorithm, Pal and Bezdek(1997) proposed a clustering technique in order to achieve a better clustering model that integrate the features of both FCMC and PCMC by using the fuzzy values of the FCMC as well as the typicality values of the PCMC. They named this clustering technique as fuzzy possibilistic c-means clustering [FPCMC]. Let us suppose that M-dimensional a set of n data points represented by x_i (i = 1, 2,...,n), are to be clustered into c number of clusters $(1 < c < n)$ and a partition matrix $U = [u_{i,j}] \in [0,1]$, $i = 1,2,...,n, j = 1,2,...,c$, where $u_{i,j}$ tells the degree to which element x_i belongs to cluster C_j. Let $C = \{c_1, c_2, ..., c_c\}$ is list of $c\{\backslash displaystyle\, c\}$ cluster centres. An objective function for FPCMC algorithm depending on both membership and typicality values is given by:

$$J_{m,\eta} = \sum_{j=1}^{c} \sum_{i=1}^{n} \left(u_{i,j}^m + t_{i,j}^\eta \right) x_i - c_j^2$$

Subject to the constraints:

$$\sum_{j=1}^{C} u_{i,j} = 1, \forall i \in \{1, 2, ..., n\}$$

$$\sum_{i=1}^{n} t_{i,j} = 1, \forall j \in \{1, 2, ..., c\}$$

where m $(m > 0)$ and $\eta(\eta > 0)$ are both weighting exponents also $0 \leq u_{i,j}, t_{i,j} \leq 1$.

Using the above constraints and conditions, the first order necessary conditions for extreme of $J_{m,\eta}$ in terms of Lagrange multiplier theorem would be as follows:

$$t_{i,j} = \cfrac{1}{\sum_{k=1}^{n} \left(\cfrac{d_{i,j}}{d_{i,k}} \right)^{\frac{2}{m-1}}}, \forall i, j$$

$$c_j = \frac{\sum_{i=1}^{n} \left(u_{i,j}^m + t_{i,j}^\eta \right) x_i}{\sum_{i=1}^{n} \left(u_{i,j}^m + t_{i,j}^\eta \right)}, \forall j$$

where $d_{i,j}$ is the distance of the data point x_i to the prototype c_j which is computed as:

$$d_{i,j} = x_i - c_j = \left(x_i - c_j \right)^T A \left(x_i - c_j \right)$$

where A is symmetric positive definite matrix. When A is identity matrix, $d_{i,j}$ represents Euclidean distance which represents the similarity between cluster center and data points. FPCMC generates possibilities and memberships at the same time for the data point.

Possibilistic Fuzzy c- Means Clustering [PFCMC] Algorithm

In FPCMC, the constraint according to which the sum of all membership values of all data points in a cluster must be equal to one may lead to a problem for big data sets. To resolve this problem Pal et al.(2005) proposed a mixture clustering model called possibilistic-fuzzy c-means clustering (PFCMC) algorithm that comprises a probabilistic and a possibilistic terms. Let, M-dimensional set of n data points represented by x_i (i = 1, 2,...,n), are to be clustered into c number of clusters (1 < c < n) and a partition matrix $U = [u_{i,j}] \in [0,1]$, $i = 1,2,...,n, j = 1,2,...,c$, where $u_{i,j}$ tells the degree to which element x_i belongs to cluster C_j. Let $C = \{c_1, c_2,...,c_c\}$ is list of $c\{\backslash displaystyle\, c\}$ cluster centres. PFCMC optimizes the following objective function:

$$J_{m,\eta} = \sum_{j=1}^{c}\sum_{i=1}^{n}\left(au_{i,j}^{m} + bt_{i,j}^{p}\right)x_i - c_j^2 + \sum_{j=1}^{c}\eta_j\sum_{i=1}^{n}\left(1 - t_{i,j}\right)^{p}$$

where the fuzzy membership functions, $u_{i,j}$ are constrained by the probabilistic conditions, while the typicality values, $t_{i,j} \in [0,1]$ are subject to:

$$0 < \sum_{j=1}^{c} t_{i,j} < c, \forall j = 1,2,...c.$$

The parameters ' a ' and ' b ' define the relative importance between the membership values and the typicality values. The fuzzy exponent m and possibilistic exponent p must be greater than 1. The variables η_j (j = 1,2,...,c) are called possibilistic penalty terms and control the variance of the clusters. The optimization formulas applied in each loop of the alternating optimization are:

$$t_{i,j}^{*} = \cfrac{1}{1 + \left(\cfrac{bx_i - c_j}{\eta_j}\right)^{\frac{1}{(P-1)}}} \quad, \forall j = 1,2,...,c, \; \forall i = 1,2,...,n$$

$$c_j^* = \frac{\sum_{i=1}^{n}\left(au_{i,j}^m + bt_{i,j}^p\right)x_i}{\sum_{i=1}^{n}\left(au_{i,j}^m + bt_{i,j}^p\right)} \quad, \forall\, j = 1,2,\dots,c$$

$$u_{i,j}^* = \frac{1}{\sum_{k=1}^{n}\left(\dfrac{x_i - c_j}{x_i - c_k}\right)^{\frac{2}{m-1}}} = \frac{x_i - c_j^{-2/(m-1)}}{\sum_{k=1}^{c} x_i - c_k^{-2/(m-1)}}$$

$$\forall i = 1,2,\dots,n \quad \forall\, j = 1,2,\dots,c$$

The probabilistic part $u_{i,j}^*$ of the partition computed exactly the same way as in FCMC algorithm.

Gustafson-Kessel Clustering [GKC] Algorithm

Gustafson and Kessel(1979) modified the FCMC algorithm. Let us suppose a set of n data points which are represented by x_i (i = 1, 2,...,n) to be clustered into c number of clusters ($1 < c < n$). The algorithm returns a list of $c\left\{\backslash displaystyle c\right\}$ cluster with centres $C = \left\{c_1, c_2, \dots, c_c\right\}$. The local adaptation of the distance metric to the shape of the cluster by estimating the cluster covariance matrix and adapting the distance-inducing matrix correspondingly is the main feature of the GKC algorithm. In this algorithm, each cluster is associated with a separate matrix A_j. The distance between data point x_i and the cluster center c_j is calculated as:

$$d_{ij}^2 = \left(x_i - c_j\right)^T A_j\left(x_i - c_j\right), \ \forall i = 1,2,\dots,n, \forall j = 1,2,\dots,c$$

The matrices A_j are used as optimization variables in the c-means functional. Thus in GKC algorithm, each cluster can adopt the distance norm to the local topological structure of the data. The objective function for GKC algorithm is defined by:

$$J = \sum_{j=1}^{C}\sum_{i=1}^{n}\left(u_{i,j}\right)^m d_{ij}^2$$

Since the objective function, J is linear in A_j therefore it cannot be directly minimized with respect to A_j. To obtain a feasible solution, A_j must be constrained in some way. This is accomplished by making constrains for the determinant of A_j as:

$$|A_j| = \rho_j, \rho_j > 0, \forall j$$

The coefficient ρ_j defines the volume of a particular cluster (if we do not have the knowledge of the problem, we may take $\rho_j = 1$). The following expression for A_j is obtained by using Lagrange multiplier method:

$$|A_j| = \left[\rho_j \det(F_j)\right]^{1/n} F_j^{-1}$$

where F_j is so called fuzzy covariance matrix for the j-th cluster which is given by:

$$F_j = \frac{\sum_{i=1}^{n} (u_{i,j})^m (x_i - c_j)(x_i - c_j)^T}{\sum_{i=1}^{n} (u_{i,j})^m}$$

The algorithm initialization requires determining the same parameters as in FCMC algorithm. GKC algorithm finds clusters of any shapes but it run time complexity is higher than the FCMC algorithm due to the necessity to compute the determinant and inverse of the matrix F_j in each iteration.

Babuska et al. (2002) have presented two techniques to improve the calculation of the fuzzy covariance matrix in the GKC algorithm. This technique reduces the risk of over fitting when the number of training samples is less in comparison to the number of data clusters. This can be achieved by adding a scaled unity matrix to the calculated covariance matrix.

Entropy-Based Fuzzy Clustering (EBFC) Algorithm

Yao et al. (2000) have presented an entropy-based fuzzy clustering algorithm. In EBFC algorithm, entropy values of each data point are calculated first and then the data point with minimum entropy value is selected as the cluster center. The data points, which are not being selected inside any of the clusters, are termed as outliers.

Consider a set X, of n data points in a M dimensional space and each data point x_i (i=1,2,...,n) is represented by a set of M values (i.e., $x_{i1}, x_{i2}, x_{i3}, \ldots x_{iM}$). Thus, the data set can be represented by an N ×M matrix. Values of each dimension are normalized in the range [0.0 - 1.0]. The Euclidean distance for each pair of data points (e.g., i and j) is determined as follows:

$$d_{ij} = \sqrt{\sum_{k=1}^{M}\left(x_{ik} - x_{jk}\right)^2}$$

An entropy value between two data points lies in the range [0.0, 1.0]. The entropy will be very low (close to 0.0) for very close or very distant pairs of data points and it will be very high (close to 1.0) for those data points separated by the distance close to the mean distance of all pairs of data points.

The total entropy value, E_i for a data point x_i with respect to all other data points is calculated as:

$$E_i = -\sum_{\substack{j \in X}}^{j \neq i}\left(S_{ij} \log_2 S_{ij} + \left(1 - S_{ij}\right)\log_2\left(1 - S_{ij}\right)\right),$$

where S_{ij} is the similarity between the data points x_i and x_j normalized to [0.0, 1.0]. During the process of clustering, the data point having the minimum entropy value is selected as the cluster center. Similarity between any two points (i.e., i and j) is calculated as follows:

$$S_{ij} = e^{-\alpha d_{ij}}$$

where α is a numerical constant. The experiments with various values for α suggest that it must be applicable mostly for all kinds of data sets, and not just for certain data sets. It is to be noted that the similarity value S_{ij} lies in the range [0.0, 1.0]. The value of α is calculated based on the assumption that the similarity value S_{ij} is set equal to 0.5, when the distance between two data points, d_{ij} becomes equal to the mean distance $\bar{d}$, which is represented as follows:

$$\bar{d} = \frac{1}{{}^nC_2}\sum_{i=1}^{n}\sum_{j>i}^{n}d_{ij}$$

From $S_{ij} = e^{-\alpha d_{ij}}$, α can be calculated as:

$$\alpha = -\frac{\ln 0.5}{\bar{d}}$$

Hence, α is determined by the data and can be calculated automatically. The clustering algorithm is explained below.

Step 1: Calculate entropy E_i (i = 1, 2, 3, ...,n) for each data point x_i of X.

Step 2: Choose x_{iMin} with least entropy E_i .

Step 3: Put x_{iMin} and the data points having similarity with x_{iMin} in greater than threshold value β for similarity in a cluster and remove this from X.

Step 4: If X is not empty go to Step 2.

Chattopadhyay et al. (2011) have compared FCMC and EBFC algorithms on four data sets (IRIS, WINES, OLITOS and Psychosis) in terms of the quality of the clusters (i.e., discrepancy factor, compactness, and distinctness).

Self-Organizing Maps Clustering (SOMC) Algorithm

A self-organizing map clustering (SOMC) is a kind of artificial neural network (ANN) that is trained by using unsupervised learning process to produce a low dimensional output. Self-organizing maps are different from the other ANN in the sense that they use a neighbourhood function to preserve the topological properties of the input space. Self-organising maps is an excellent tool for making the clusters of data points. The architecture of the SOMC was introduced by T. Kohonen (1982, 1995) with examples. Adapting to the training set SOMC forms its outputs by itself. The basic SOMC consists of k neurons located on a regular low-dimensional lattice with vectors of n input signals. The lattice is usually either hexagonal or rectangular. Measure of distance is used to determined similarity between the input vector and neurons. Some popular distances among input pattern and SOMC units are:

1.   Euclidian distance
2.   Squared Euclidian distance
3.   Direction cosine distance
4.   Chebychev distance
5.   Pearson distance

SOMC has been widely used in many industrial areas such as data compression, pattern recognition, biological modelling, data mining and signal processing. The stages of the SOMC algorithm are summarized as follows:

Step 1: Fix the number of output neurons k, each of which represent a cluster.
Step 2: Initialization- Choose random values for each initial weight vector w_i.
Step 3: Sampling- Calculate sample training input vector, X from the input space.
Step 4: Set topological neighborhood parameters and learning rate parameter α.
Step 5: For each input X, update weight vector w_j, of the closest output neurons and its neighbors by applying the following weight update equation

$$w_{ji}(new) = w_{ji}(old) + \alpha\left[X_i - w_{ji}(old)\right], \ i=1,2,3,...n.$$

Step 6: Repeat steps 2 to 5 until the feature map stops changing.

For the stability to be guaranteed and convergence, the learning rate α and neighbourhood radius are decreasing with each iteration towards the zero.

Some Examples on Clustering Algorithms

Example-1: To explain the concept of fuzzy c mean algorithm, a set $X = \{p_1, p_2,..., p_7\}$ of seven processors of distributed communication system (DCS) has been considered. The groups of processors are clustered together based on the architecture of DCS and the application demand. The Inter processor distance between each pair of processor has been shown in the following table.

processors →↓	p_1	p_2	p_3	p_4	p_5	p_6	p_7
p_1	0	11	18	17	15	14	19
p_2	11	0	19	18	14	16	17
p_3	18	19	0	13	10	15	17
p_4	17	18	13	0	19	12	11
p_5	15	14	10	19	0	18	16
p_6	14	16	15	12	18	0	10
p_7	19	17	17	11	16	10	0

Let us consider the fuzziness coefficient m = 1.25 and number of cluster c = 3. Here $U = [u_{i,j}] \in [0,1]$, $i = 1, 2, ..., 7, j = 1, 2, 3$ represent the partition matrix.

Let us consider initial $U = [u_{i,j}]$ as:

$$U = \begin{array}{c} \\ C_1 \\ C_2 \\ C_3 \end{array} \begin{array}{ccccccc} P_1 & P_2 & P_3 & P_4 & P_5 & P_6 & P_7 \\ 0.2 & 0.1 & 0.8 & 0.1 & 0.8 & 0.1 & 0.0 \\ 0.1 & 0.2 & 0.1 & 0.7 & 0.1 & 0.7 & 0.8 \\ 0.7 & 0.7 & 0.1 & 0.2 & 0.1 & 0.2 & 0.2 \end{array}$$

Calculated cluster centers are:

$$c_1 = \begin{bmatrix} 27.33 & 28.35 & 12.62 & 28.17 & 12.44 & 28.41 & 29.64 \end{bmatrix}$$
$$c_2 = \begin{bmatrix} 37.56 & 37.11 & 34.91 & 21.17 & 39.07 & 20.03 & 18.64 \end{bmatrix}$$
$$c_3 = \begin{bmatrix} 14.50 & 14.50 & 29.10 & 26.35 & 24.83 & 24.00 & 26.93 \end{bmatrix}$$

Now updating partition matrix U according to step 4 is:

$$U = \begin{array}{c} \\ C_1 \\ C_2 \\ C_3 \end{array} \begin{array}{ccccccc} P_1 & P_2 & P_3 & P_4 & P_5 & P_6 & P_7 \\ 0.44 & 0.43 & 0.70 & 0.52 & 0.69 & 0.48 & 0.5 \\ 0.34 & 0.21 & 0.29 & 0.47 & 0.22 & 0.44 & 0.5 \\ 0.22 & 0.36 & 0.01 & 0.01 & 0.09 & 0.08 & 0.0 \end{array}$$

Repeating the steps 3-6, the optimal partition matrix U, will be:

$$U = \begin{array}{c} \\ C_1 \\ C_2 \\ C_3 \end{array} \begin{array}{ccccccc} P_1 & P_2 & P_3 & P_4 & P_5 & P_6 & P_7 \\ 0.09 & 0.09 & 0.85 & 0.16 & 0.73 & 0.12 & 0.0 \\ 0.08 & 0.08 & 0.09 & 0.71 & 0.11 & 0.74 & 0.88 \\ 0.83 & 0.83 & 0.06 & 0.13 & 0.16 & 0.14 & 0.22 \end{array}$$

The output of fuzzy c mean algorithm comprises of a partitioning of the data with assignment of membership value to data points to each clusters. Higher membership value of a data point in a cluster indicates its greater relevance to that cluster. On the basis of membership values of each processor in the optimal partition matrix U, the following three clusters could be as:

$$C_1 = \{p_3, p_5\}, C_2 = \{p_4, p_6, p_7\} \text{ and } C_3 = \{p_1, p_2\}$$

Example-2: To illustrate the fuzzy k mean clustering algorithm, consider the following data set X={ x_1, x_2, x_3, x_4, x_5 } consisting of the scores of two variables on each of five individuals.

Now we want to make two groups (two clusters) of this data points. Here k=2. We assume the initial partition that classifies the data set into clusters $C_1 = \{x_1, x_2\}$ and $C_2 = \{x_3, x_4, x_5\}$. The clusters centroids are:

$$c_1 = (2, 3.5)$$

$$c_2 = (4, 4.3)$$

Distances of each data point from each of cluster centroid are given in the Table 3. Update the cluster by assigning the data points to the nearest cluster centroid. The new clusters are $C_1 = \{x_1, x_3\}$ and $C_2 = \{x_2, x_4, x_5\}$. Updated clusters centroids are:

$$c_1 = (1.5, 1.5)$$

$$c_2 = (4.3, 5.7)$$

Table 2.

Data Points	A	B
x_1	1	2
x_2	3	5
x_3	2	1
x_4	4	5
x_5	6	7

Distances of each data point from each of the new cluster centroid are given in the Table 4.

Update the cluster by assigning the data points to the nearest cluster centroid. The new clusters are $C_1 = \left\{ x_1, x_3 \right\}$ and $C_2 = \left\{ x_2, x_4, x_5 \right\}$, which are the same as previous clusters. So these are the final clusters. Thus the clusters of the five data points are:

$$C_1 = \left\{ x_1, x_3 \right\} \text{ and } C_2 = \left\{ x_2, x_4, x_5 \right\}$$

Clusters C_1 and C_2 are also depicted by Figure 1.

APPLICATIONS OF CLUSTERING ALGORITHM

New technology has generated more complex and challenging tasks, requiring more powerful clustering. Clustering algorithms can be applied in many fields, for instance:

1.   Clustering Algorithm in Search Engines
2.   Clustering Algorithm in Drug Activity Prediction
3.   Clustering Algorithm in Biology
4.   Clustering Algorithm in Academics

Table 3.

Data Points	Cluster Centroid	
	c_1	c_2
x_1	1.8	3.8
x_2	1.8	1.22
x_3	2.5	3.8
x_4	2.5	0.7
x_5	5.3	3.36

Table 4.

Data Points	Cluster Centroid	
	c_1	c_2
x_1	0.5	4.9
x_2	3.8	1.48
x_3	0.5	4.88
x_4	4.3	0.7
x_5	7.1	2.14

5. Clustering Algorithm in Data Mining
6. Business and marketing:
7. Computer science: Clustering algorithms are widely used in computer science in the following applications:
   a. Software evolution
   b. Image segmentation
   c. Evolutionary algorithms
   d. Markov chain Monte Carlo methods
   e. Anomaly detection
   f. Recommender systems

Figure 1. Graphical representation of clusters

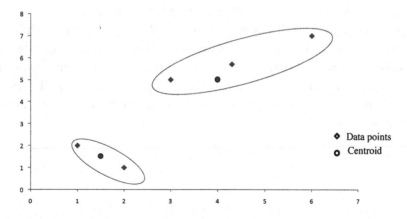

g.   Wireless Sensor network's based application
h.   Field robotics
i.   Image processing
j.   Voice mining
k.   Pattern recognition
l.   Machine learning
m.   Sequence analysis
n.   Earthquake studies
o.   Petroleum geology
p.   Mathematical chemistry

8.   **Social Science:** Some applications of clustering algorithms where the clustering have been used in field of social science are:

a.   Crime analysis
b.   Educational data mining
c.   Physical geography
d.   Climatology
e.   City-planning
f.   Libraries
g.   Insurance
h.   Whether report analysis

CONCLUSION

Clustering is an important technique in data analysis and data mining applications. It divides data into groups of similar objects. This chapter starts at the basic definitions of clustering and lists the commonly used distance (dissimilarity) functions. The chapter describes different methodologies and parameters associated with different clustering algorithms. A number of important well-known clustering methods from literature have been discussed. Their advantages and disadvantages over earlier existing methods have been discussed. It is impossible to develop a general clustering algorithm including all cases. This is because in practical all clustering algorithms possess their own characteristics and limitations. In the context of each application, some clustering algorithms have seen more appropriate than other. However, the issue of choosing a proper clustering algorithm for a particular problem is still a subject of research.

REFERENCES

Anderberg, M. R. (1773). Cluster analysis for applications. Academic Press.

Babuska, R. V., Vander, P. J., & Kaymak, U. (2002). Improved Covariance Estimation for Gustafson-Kessel Clustering. *Proceedings of the 2002 IEEE international conference on Fuzzy Systems*, 1081-1085.

Bezdek, J. C. (1983). *Pattern Recognition with Fuzzy Objective Function Algorithms*. New York: Plenum.

Chattopadhyay, S., Pratihar, D. K., & Sarkar, S. C. D. (2011). A Comparative Study of Fuzzy c-Means Algorithm and Entropy-Based Fuzzy Clustering Algorithms. *Computer Information*, *30*, 701–720.

Dunn, J. C. (1773). A Fuzzy Relative of the ISODATA Process and its Use in Detecting Compact Well-Separated Clusters. *Journal of Cybernetics, Vol.*, *3*(3), 32–57. doi:10.1080/01969727308546046

Fraley, C., & Raftery, A. (1999). MCLUST: Software for Model-Based Cluster Analysis. *J. Classificat.*, *16*(2), 297–306. doi:10.1007003579900058

Genther, H., Runkler, T. A., & Glesner, M. (1994). Defuzzification based on fuzzy clustering. *Proceeding of IEEE World Congress on Computational Intelligence*.

Gustafson, D. E., & Kessel, W. C. (1978). Fuzzy Clustering with a Fuzzy Covariance Matrix. *IEEE Conference on Adaptive Processes*, *17*(1), 761-766.

Han, J., & Kamber, M. (2001). *Data Mining*. San Francisco, CA: Morgan Kaufmann Publishers.

Hathaway, R., Bezdek, J., & Hu, Y. (2000). Generalized fuzzy -means clustering strategies using norm distances. *IEEE Transactions on Fuzzy Systems*, *8*(5), 576–582. doi:10.1109/91.873580

Jain, A. K., & Dubes, R. C. (1988). *Algorithms for Clustering Data*. Englewood Cliffs, NJ: Prentice Hall.

Jeffrey, D. B., & Raftery, A. E. (1993). Model-Based Gaussian and Non-Gaussian Clustering. *Biometrics*, *49*(3), 803–821. doi:10.2307/2532201

Jin, X., & Han, J. (2010). K-Medoids Clustering. In *Encyclopedia of Machine Learning* (pp. 564–565). Springer. doi:10.1007/978-0-387-30164-8_426

Kaufman, L., & Rousseeuw, P. J. (1987). *Clustering by Means of Medoids in Statistical Data Analysis Based on the L_1 - Norm and Related Methods*. North- Holland.

Kohonen, T. (1982). Self-organized formation of topologically correct feature maps. *Biological Cybernetics*, *43*(1), 59–69. doi:10.1007/BF00337288

Kohonen, T. (1995). *Self-Organizing Maps* (Vol. 30). Berlin: Springer. doi:10.1007/978-3-642-97610-0

Krishnapuram, R., & Keller, J. (1993). A Possibilistic Approach to Clustering. *IEEE Transactions on Fuzzy Systems*, *1*(2), 98–110. doi:10.1109/91.227387

Kumar, H. (2017). Some Recent Defuzzification Methods. In *Theoretical and Practical Advancements for Fuzzy System Integration* (pp. 31–48). IGI-Global. doi:10.4018/978-1-5225-1848-8.ch002

Leung, Y., Zhang, J., & Xu, Z. (2000). Clustering by scale-space filtering. *IEEE Transactions on Pattern Analysis and Machine Intelligence*, *22*(12), 1396–1410. doi:10.1109/34.895974

Lloyd, S. (1982). Least Squares Quantization in PCM. *IEEE Transactions on Information Theory*, *28*(2), 129–137. doi:10.1109/TIT.1982.1056489

MacQueen, J. B. (1967). Some Methods for Classification and Analysis of Multivariate Observations. In *Proceedings of the 5th Berkeley Symposium on Mathematical Statistics and Probability*. Berkeley, CA: University of California Press.

Melnykov, V., & Maitra, R. (2010). Finite Mixture Models and Model-Based Clustering. *Statistics Surveys*, *4*(0), 80–116. doi:10.1214/09-SS053

Morzy, T., Wojciechowski, M., & Zakrzewicz, M. (1999). Pattern-Oriented Hierarchical Clustering. *Proceeding of 3rd East Eur. Conf. Advances in Databases and Information Systems,* 179–190. 10.1007/3-540-48252-0_14

Pal, N. R., Pal, K., & Bezdek, J. C. (1997). A Mixed c-Means Clustering Model. In *Proceedings of the 6th IEEE International Conference on Fuzzy Systems* (Vol. 1, p. 11). IEEE. 10.1109/FUZZY.1997.616338

Pal, N. R., Pal, K., Keller, J. M., & Bezdek, J. C. (2005). A Possibilistic Fuzzy c-Means Clustering Algorithm. *IEEE Transactions on Fuzzy Systems*, *13*(4), 517–530. doi:10.1109/TFUZZ.2004.840099

Park, H. S., & Jun, C. H. (2009). A Simple and Fast Algorithm for K-Medoids Clustering. *Expert Systems with Applications*, *36*(2), 3336–3341. doi:10.1016/j.eswa.2008.01.039

Selim & Ismail. (1984). K-means-Type Algorithms: A Generalized Convergence Theorem and Characterization of Local Optimality. *IEEE Transactions on Pattern Analysis and Machine Learning, 6*(1), 81-87.

Steinhaus, H. (1957). Sur la Division Des Corps Matériels en Parties. *Bull. Acad. Polon. Sci., 4*(12), 801–804.

Xu, R., & Wunsch, D. II. (2005). Survey of Clustering Algorithms. *IEEE Transactions on Neural Networks, 16*(,3), 645–677. doi:10.1109/TNN.2005.845141 PMID:15940994

Yao, J., Dash, M., Tan, S. T., & Liu, H. (2000). Entropy-Based Fuzzy Clustering and Fuzzy Modeling. *Fuzzy Sets and Systems, 113*(3), 381–388. doi:10.1016/S0165-0114(98)00038-4

Zahn, C. T. (1971). Graph-Theoretical Methods for Detecting and Describing Gestalt Clusters. *IEEE Transactions on Computers, 20*(1), 68–86. doi:10.1109/T-C.1971.223083

Chapter 10
Addressing Security Issues and Standards in Internet of Things

Sushruta Mishra
KIIT University, India

Soumya Sahoo
C. V. Raman College of Engineering, India

Brojo Mishra
C. V. Raman College of Engineering, India

ABSTRACT

In the IoTs era, the short-range mobile transceivers will be implanted in a variety of daily requirements. In this chapter, a detail survey in several security and privacy concerns related to internet of things (IoTs) by defining some open challenges are discussed. The privacy and security implications of such an evolution should be carefully considered to the promising technology. The protection of data and privacy of users has been identified as one of the key challenges in the IoT. In this chapter, the authors present internet of things with architecture and design goals. They survey security and privacy concerns at different layers in IoTs. In addition, they identify several open issues related to the security and privacy that need to be addressed by research community to make a secure and trusted platform for the delivery of future internet of things. The authors also discuss applications of IoTs in real life. A novel approach based on cognitive IoT is presented, and a detailed study is undertaken. In the future, research on the IoTs will remain a hot issue.

DOI: 10.4018/978-1-5225-5793-7.ch010

INTRODUCTION

With the rapid development of Internet technology and communications technology, our lives are gradually led into an imaginary space of virtual world. People can chat, work, shopping, keeps pets and plants in the virtual world provided by the network. However, human beings live in a real world; human activities cannot be fully implemented through the services in the imaginary space. It is the limitation of imaginary space that restricts the development of Internet to provide better services. To remove these constraints, a new technology is required to integrate imaginary space and real-world on a same platform which is called as Internet of Things (IoTs). Internet of Things (IoT) represents the extension and evolution of the Internet, which has great potential and prospects for modern intelligent service and applications. Based on a large number of low-cost sensors and wireless communication, the sensor network technology puts forward new demands to the Internet technology. It will bring huge changes to the future society, change our way of life and business models. The ability to connect, communicate with, and remotely manage an incalculable number of networked, automated devices via the Internet is becoming pervasive, from the factory floor to the hospital operating room to the residential basement. The transition from closed networks to enterprise IT networks to the public Internet is accelerating at an alarming pace and justly raising alarms about security. In the coming day's world will witness a massive surge of Internet. In his article (Rossi 2015). it is predicted that by 2020 more than 20 billion devices will be connected through Internet. Cisco has predicted about 50 billion devices will be associated with Internetworking (Evans 2011). The, term Internet of Things (IoT) has first been used by Bill Gates in 1995 (Gates 1995). The IoT technology draws huge changes in everyone's everyday life. In the IoTs era, the short-range mobile transceivers will be implanted in variety of daily requirements. The connections between people and communications of people will grow and between objects to objects at any time, in any location. The efficiency of information management and communications will arise to a new high level (ITU 2005). The dynamic environment of IoTs introduces unseen opportunities for communication, which are going to change the perception of computing and networking. As we become increasingly reliant on intelligent, interconnected devices in every aspect of our lives, security becomes an important issue that need to be addressed. The definition and scope of IoT have changed and its coverage has been greatly expanded (Liang et. al 2011). Now, the IoT is proverbially applied in the field of modern service, such as healthcare, smart environments, personal and social, entertainment, transportation and logistics (Atzori et.al 2010; Liang et al. 2011; Tong, Ban 2014,) .Multiple attacks have already been reported in the past against different embedded devices (Bansal 2014; Wright 2011) and we can expect many more in the IoT domain. Security is one of the foremost concerns raised by different stakeholders in Internet of Things which has the potential to slow down its adoption. While security has always been a concern since the computers started connecting to each other, the impact was limited to stealing money and intellectual property. But Internet of things adds a completely new dimension to this where the devices performing critical tasks, if insecure, can be manipulated to the devastating effects. The impact could be on public safety, environment, productivity and many others. Apart from benefits of IoTs, there are several security and privacy concerns at different layers. The privacy and security implications of such an evolution should be carefully considered to the promising technology. The protection of data and privacy of users has been identified as one of the key challenges in the IoT relies on the principle of the extensive processing of data through sensors that are designed to communicate unobtrusively and exchange data in a seamless way; the exponential volume of data that can be collected, and its further combination, its storage in the cloud and the use of predictive analytics tools can transform data into something useful but also allow companies - and potentially malware - to have very detailed profiles of individuals; and the sharing and combination of data through cloud services will increase the locations and jurisdictions where personal data resides.

MOTIVATION

With the rapid advances in low-cost wireless sensors, radio frequency identification (RFID), Web technologies and wireless communications recently, connecting various smart objects to Internet and realizing the communications of machine-to-human and machine-to-machine with the physical world have been expected widely. That is the concept of Internet of Things (IoT), which can provide ubiquitous connectivity, information gathering and data transmitting capabilities in different fields, such as health monitoring, emergencies, environment control, military and industries. The pervasive sensing and control capabilities brought by IoT will change our daily life significantly]. In an era of IoT, there are billions of devices linked to the Internet. Such a large number of devices deployed in the IoT lead to many technical challenges including spectrum scarcity, energy consumption and security [With the rapid advances in low-cost wireless sensors, radio frequency identification (RFID), Web technologies and wireless communications recently, connecting various smart objects to Internet and realizing the communications of machine-to-human and machine-to-machine with the physical world have been expected widely .That is the concept of Internet of Things (IoT), which can provide ubiquitous connectivity, information gathering and data transmitting capabilities in different fields, such as health monitoring, emergencies, environment control, military and industries. The pervasive sensing and control capabilities brought by IoT will change our daily life significantly. The early visionaries of the Internet of Things, foresaw a time when practically any physical object could be equipped with sensors and hooked up to the Internet to translate the physical world into digital information. Therefore now the focus lies on the factory assembly lines, electrical grids, automobiles, highways, buildings, and the like. The goal was to gather streams of information from sensors that could be used to automate processes—such as balancing supply and demand in a power grid—and operate more efficiently. Today, these applications have become mainstream, so it's time to take advantage of a second generation of Internet of Things technologies and capabilities called as the Cognitive Internet of Things. Cognitive IoT technologies will make it possible for business leaders to understand what's happening in the world more deeply. By infusing intelligence into systems and processes, businesses will be able to not only do things more efficiently, but to improve customer satisfaction, to discover new business opportunities, and to anticipate risks and threats so they can better deal with them. As powerful as today's sensor networks are, they aren't up to the task of unlocking the complex interrelationships between people, places and things that drive business and the economy. To reach the next level, businesses need cognitive technologies that enable them to gather and integrate data from many types of sensors and other sources, to reason over that data, and to learn from their interactions with it. Think of it this way: First-generation IoT technologies gave us nuggets of information that could make a big difference in achieving operational efficiencies. The next generation creates vast communities of devices that share information, which in turn can be interpreted in a larger context and managed by people using cognitive systems. In the era of Cognitive IoT, no machine is an island. Today, focus is on what it will take to reap the full benefits of the Internet of Things—the addition of cognitive technologies. It is a global "movement" that billions of people will benefit from and many organizations will help propel forward. Now IBM has a broad portfolio of technologies for managing the data gathered from sensors. They were working with the UK's utility provider, National Grid, to proactively maintain the health of the grid, keeping the lights on across Britain and also with Vodafone in Spain to analyze information from sensors in cities—energy, water, emergency management, healthcare—to improve operations and the quality of life. These days, about 90 percent of the data that's gathered by sensors is lost or thrown away for a variety of reasons—including bandwidth limitations and constraints driven by security and privacy. In addition, a wealth of unstructured data is available from sources ranging from news Web sites to call centers to social networks. With the new cognitive IOT capabilities and Watson, our clients can combine all of these diverse sources of data in real time, understand what's going on more deeply, and derive valuable insights.

There was the addition of several cognitive technologies to our IoT portfolio—machine learning, natural language processing, video and image analytics and text analytics. Moreover these things can be tapped into these capabilities in the cloud to enhance existing IoT-based applications and build new ones. More cognitive capabilities will be added in the coming months. In addition, working with leaders of industries, we're creating industry-specific solutions that combine IoT and cognitive—starting with telecommunications, real estate, aerospace and retailing. For decades, IBM has worked with some of the largest banks, financial institutions, insurance companies and others protecting security of these clients. We have 48 cloud data centers across the globe giving clients the choice of locating their data where they want it. To give you a clearer picture of the shift that's underway, let me lay out a scenario. Imagine a large department store. Today, facilities people tap networks of sensors to better manage the temperature and energy use. Tomorrow, thanks to cognitive technologies, the store, essentially, becomes self aware.

The store bristles with embedded technology. It's blanketed with unobtrusive video cameras. Sensors on merchandise and shelves are hooked up to the network—as are shoppers who have agreed to be connected via smart phone apps. Audio speakers facilitate two-way conversations with people and gather information about what's being said. A cognitive system combines all of this sensor data with information gathered about the local weather and news, social networking streams, and sales trends. These new capabilities enable managers to understand what's going on in the store in real time, interact with shoppers, and anticipate changes. If it's raining outside, digital signs in the store might direct shoppers to umbrellas, rain gear or hair-care products. Video analytics tools discover the demographics of people buying certain items. If a large number of shoppers pick up an item but don't buy it, machine learning algorithms spot patterns that signal what's going wrong. Stores will be able to provide shoppers with cognitive assistants—via their smart phones—that know them and provide them with superior in-store experiences. There are a host of other situations where Cognitive IoT can make a big difference. For instance, an airline might combine data from sensors measuring stress on aircraft with turbulence data to optimize maintenance schedules—potentially heading off expense repairs, or God forbid, a failure in flight.

BACKGROUND

As stated above, architecture is the basis for a new kind of network, which guides the technology activities and engineering practices. Thus, researches orienting to IoT architecture have been widely concerned. Through analyzing the technical framework of the Internet and the logical layer architecture of the Telecommunication Management Network, a new five-layer architecture of IoT was established which is helpful to understand the essence of the IoT (Wright 2011). Combining with the

EPC global standards, a working model of three-layer for IoT was presented, which consist of the sensing layer, network layer and application layer. Some pressing issues were disused to further promote the practical application of the IoT's architecture, such as the smart nodes optimization, information security and data processing (Zhang, Zhu 2011). In his article (Kaswar et al. 2010) had discussed about a hierarchy of architecture with increasing levels of real-world awareness and interactivity is introduced. In particular, the activity-, policy-, and process aware smart objects are described and how the respective architectural abstractions support increasingly complex application is demonstrated. In those literatures, the architectures of IoT were constructed utilizing traditional layered/hierarchic approach which is applied to Internet. No matter three-layer or five-layer, the diversity between IoT and Internet has not been deep considered in those architectures. So, the architectures cannot meet the applications of IoT ideally. Some researchers devote to establish architectures of IoT based on the Social Network. Through mapping information, objects and persons which are the major macro elements of human society to Internet, IoT and Social Network, a platform was proposed to cluster the Internet, IoT and Social Network together and promote the development of IoT (Ding et al. 2010) . Based on the notion of social relationships among objects, a novel paradigm of Social Internet of Things (SIoT) was proposed and a preliminary architecture for the implementation of SIoT was presented (Atzori et al. 2011). In their article (Ning, Wang 2011), had focused that the future IoT architecture would be considered in two aspects: Unit IoT and Ubiquitous IoT. Ubiquitous IoT refers to the global IoT or the integration of multiple Unit IoT with ubiquitous characters. Oriented to special applications, the architecture of Unit IoT was built from man like neural network model. And the architecture of Ubiquitous IoT employs the social organization framework model. Those architectures based on the Social Network are helpful to interpret the relationship between IoT and reality world, but there are still many problems to be resolved before applied.

IoT can be considered as an application extension of Internet. So, the architectures based on Internet look like more reasonable and are paid attention by many researchers. In order to integrate real-world devices to Web, the possible integration methods, in particular how the Representational State Transfer (REST) principles can be applied to embedded devices, were discussed and applied to Sun SPOT platform (Guinard 2009)[. In their article (Guinard et al. 2011) had suggested about a Web of Things (WoT) architecture and best practices based on the REST principles were described. It makes smart things become an integral part of Web and easier to build upon. Web IoT, a novel web application framework based on Google Web Toolkit, provides users with simple methods for integrating smart things, and enhances the interaction between humans and things or among things(Castellani 2012) . A novel architecture of Sensor Networks for All-IP World (SNAIL) includes a complete IP adaptation method and four significant network protocols: mobility, web enablement, time synchronization, and security. The feasibility and interoperability of the proposed approach were confirmed by the implementation of SNAIL platform(Hong 2010) .

KEY PRIVACY CHALLENGES OF IOT

The Internet of Things (IoT) will enable objects to become active participants of everyday activities. Introducing objects into the control processes of complex systems makes IoT security very difficult to address. Indeed, the Internet of Things is a complex paradigm in which people interact with the technological ecosystem based on smart objects through complex processes. The interactions of these four IoT components, person, intelligent object, technological ecosystem, and process, highlight a systemic and cognitive dimension within security of the IoT. The interaction of people with the technological ecosystem requires the protection of their privacy. Similarly, their interaction with control processes requires the guarantee of their safety. Processes must ensure their reliability and realize the objectives for which they are designed. As the IoT deals with a huge number of things and their relevant data, many security challenges have to be addressed. This is true especially when things need to interact with each other across other set of things, through many security techniques and according to different policy requirements (Sundmaeker et al. 2010) .We believe that the move towards a greater autonomy for objects will bring the security of technologies and processes and the privacy of individuals into sharper focus. Furthermore, in parallel with the increasing autonomy of objects to perceive and act on the environment, IoT security should move towards a greater autonomy in perceiving threats and reacting to attacks, based on a cognitive and systemic approach. Hence, we need to bake security in, right from planning to design through to the implementation and monitoring phase, taking a risk based approach. This begins with identifying the assets/components and value of those components, because we cannot effectively protect something which we do not know it exists. Having said that, we can break-down IoT into various components per Table 1.

However, before discussing security considerations, let us appreciate some of the security challenges and risks that are posed by the advent of IoT.

The privacy and data protection challenges related to IoT that are identified are:

- **Lack of Control and Information Asymmetry:** Interaction between objects that communicate automatically and by default, between objects and individuals´ devices, between individuals and other objects, and between objects and back-end systems, which will result in the generation of data

Table 1. IoT components breakdown. This shows a figure consisting of different components of IOT.

Device or Equipment	Physical devices, endpoints e.g. sensors, ECU's, smart meters, washing machines, etc. get connected to other devices, endpoints across networks to collect/provide information about themselves and their associated environment.
Gateway or Hub	Enables these devices get equipped to connect to the outer world via ethernet, RFID, wireless, bluetooth, etc.
Network or Transport channels	Facilitates the connectivity and transmission of information from devices/gateways, e.g. IP network, GSM/CDMA, satellite networks, etc.
Facilitation	Provides the ability for the devices to send data/information across gateways/network for further storage, processing, analysis e.g. cloud computing, big data, etc.
Consumerization or Application	Allows end user/customers to consume such information on to their smart devices like tablets, smart phones/televisions, laptops, etc.

Table 2. IoT security risks This figure demonstrates about IoT security risks

The attack surface has increased	Extensive leverage of open networks e.g. internet, public cloud, etc. Sensors. Web application. USB, Wireless, Bluetooth, Zigbee, GSM, etc.
Legacy systems (out of date OS/software) no longer supported by OEMs	Software updates, security patches mostly become a forgotten concept on legacy devices especially where vendor no longer provides support making them entry points for sabotaging customer networks leading to DOS attacks, malware infliction points, ransomware, etc.
Unidentified, unauthorized and invalidated devices	Unique identification of user, devices; authentication and access control of devices which may not have an OEM supplied unique ID - these could lead to identity spoofing, phishing, rogue devices, impersonation, etc.
Unauthorized remote access	Remote diagnostics/monitoring, remote maintenance of devices, equipment carries the risk of interception and tampering, in not using secure communication thereby leading to MITM (man in the middle) attacks.
Sensitive data exposure	Sensitive or personal information like patient data on EHR/EMR if they are connected to ECG, ventilator, etc., GPS location of a vehicle to target a person, etc., - sniffing, eavesdropping, waylaying
Extensive dependence on software and applications	Most of the attacks are targeted towards application, especially web application - Injections, XSS, CSRF etc.

flows that can hardly be managed with the traditional tools used to ensure the adequate protection of the data subjects' interests and rights.

- **Quality of the User's Consent:** The possibility of rejecting certain services is not a real alternative in IoT and classic mechanisms used to obtain consent may be difficult to apply, therefore, new ways of obtaining the user's valid consent should be considered.
- **Inferences Derived From Data and Repurposing of Original Processing:** Secondary uses of data, inferences from "raw" information, sensor fusion,

make important that at each level IoT stakeholders make sure that the data is used for purposes that are compatible with the original purpose of the processing and that those purposes are known by the user

- **Intrusive Identification of Behavior Patterns and Profiling:** Generating knowledge from trivial or even anonymous data will be made easy by the proliferation of sensors and that might enable very detailed and comprehensive life and behavior patterns)

- **Limitations on the Possibility of Remaining Anonymous When Using Services Security Risks:** Weak points can occur not only at device level but also in the communication links, storage infrastructure and other inputs of this ecosystem.

In recent years, cognition and cooperation have become popular research focuses. Since Doctor Mitola presented the concept of cognitive radio (Mitola, Maguire 1999), cognitive radio network(Urgaonkar 2008,Dall et al.2011), and cognitive network(Thomas 2006, Fortuna et.al 2009, Rabbachin 2011) have greatly interested the researchers, and large numbers of achievements have been attained, which greatly promoted the evolution of network intelligence. In those researches, the cooperative thought was often adopted to address intelligence and performance for asynchronous network (Zhang 2010) [, multi-user network (Wang et.al 2009), multi agent network (Yoshino et. al 2009), autonomous multi-hop network s (Kamhoua et al. 2010), bioinspired network (Zheng et.al 2010), autonomic computing system (Zheng et.al 2011, Zhang et al.2012), and other networks(Senthil et.al2011). Besides, cross-layer design (Srivastava and M. Motani 2005) and game theory (Charilas, Panagopoulos 2010) were introduced to improve efficiency and optimize performance.

SOME MORE OPEN CHALLENGES

The IoT has considerable potential to contribute significantly to disaster resilience of communication networks as we discussed in Section III. However, prior to succeeding in the 'grand challenges', the IoT is challenged by a variety of open questions and unsolved problems. Most challenges do not arise from the lack of existing technologies, but rather from a premature development of existing technologies and in particular from a lack of common standards and deployments that seamlessly interconnect. In the following section we will point out the areas where the most pressing issues arise.

Physical Connectivity and Hardware Limitations

Physical connectivity on a hardware level is the essential foundation to enable communication between devices. Sharing the same PHY and link layer is a requirement for data exchange between neighboring devices. For the IoT this means the use of common interface cards that use the same radio frequencies, modulations, link layer technology etc. Multiply connected gateways are required for transitioning network technologies.

A large heterogeneity of network access technologies, though, not only increases complexity of inter-networking, but may also lead to severe deployment problems in the wireless domain. Various radios that consume interfering frequencies of the limited spectrum by incompatible technologies may harm communication capacities at large without an ability to mutually coordinate. In disaster scenarios, all available devices should form a single, largely connected network—as redundant as possible. Assuming the infrastructure is down (e.g., because of power blackout or cable damage), some battery-powered devices with multiple interfaces using different radio technologies such as smart-phones, tablets, laptops will have to play the role of border routers to enable physical connectivity. However, it is noteworthy that these consumer devices typically neither have a IEEE 802.15.4 nor a BLE interface, which may lead to network partitioning because sensor networks using this link layer technology are unable to interconnect at the physical layer.

Logical Network Connectivity

The aim and outstanding success of today's Internet builds on its efficient and seamless way of interconnecting networks that use heterogeneous link layer technologies. This was achieved at large scale by using IP (the Internet Protocol) as the unique networking protocol and TCP/UDP at the transport layer. Wireless sensors and other constrained wireless devices are however too often based on proprietary network stacks (e.g., Zigbee or Nordic's Speedburst) that cannot interoperate across link layers or network borders. These confined network-ing solutions typically rely on specialized gateways to connect devices with the IP-based networks (i.e., the Internet). Recently, the situation has improved, though, as the IETF has published relevant standards for the IoT. 6LoWPAN de-fines a lightweight network sublayer that enables constrained nodes (e.g., wireless sensors using IEEE 802.15.4) to interoperate natively with IPv6. 6LoWPAN thus enables a substantial fraction of IoT devices to connect directly to the Internet. It is projected that in the near future, proprietary network stacks will be phased out in favor of an IPv6 network stack using

6LoWPAN, as this brings not only benefits for vendors through standardization but also through faster time-to-market, cheaper development cycles leveraging well-known development prac-tices and tools.

Prioritization of Data Traffic

Largely heterogeneous link transitions bear the problem of exhausting congestions that are likely to kill data flows. Assuming the connectivity gap is bridged at the MAC/PHY layer and at the network layer as described in Sections IV-A and IV-B, throughput may be very limited. The general idea is to use the available throughput for the most important services. A challenge that remains is thus the design of mechanisms that guarantees that only these services do use of the available throughput.

An idea could be to introduce a 'disaster mode' for IoT devices. Besides their normal mode of operation, IoT nodes could switch to an alternative mode of operation in which the goal becomes spontaneous maximization of connectivity in the sense described in Section III. Furthermore, this special mode of operation could implement prioritization policies that would guarantee first responders or official organizations privileged access to the newly spawned communication network. This 'disaster mode' would be roughly comparable to the emergency call mode in today's mobile phones, where 911 calls are possible even if no registered SIM-card is activated.

Social Acceptance

As described in Section III, leveraging IoT devices to mitigate the impact of a disaster on network connectivity implies that devices may be required to be operated outside their intended scope, and connect to external parties that normally do not have access to those devices. For example, if privately owned sensor networks were required to relay communication traffic on behalf of governmental agencies, or on behalf of other private individuals that must send/receive emergency information, the owners of such networks would need to allow a mode of operation that they do not fully control. Social acceptance of this category of usage should be studied, to prevent situations where owners of devices actively try to block any use outside their full control—preventing in effect the approach towards more resilience.

Network Security Aspects

The IoT in general presents a number of challenges in terms of application layer and network layer security. These security challenges naturally transfer to IoT use in case of disaster scenarios. In this context, one should avoid the usual reflex of initially leaving security aspects out of the picture because "every bit of the scarce throughput should be used for communication traffic". For example, there are a number of scenarios in which unprotected network traffic could be used by malicious third parties to intentionally interrupt or alter information that is exchanged between first responders or coming from emergency calls, e.g., large-scale terrorist attacks such as 9/11. As data is routed through the IoT, attackers could try to tamper with communications ways that cripple helper organization. Furthermore, the mechanisms that trigger devices to switch to 'disaster mode' operation should itself be secure in order to prevent attacks aiming to disrupt normal network operation. These challenges are directly related to lightweight, decentralized authentication schemes.

Towards Disaster Resilience

With the technology available today, the Internet of Things cannot yet be used to improve our communication networks resilience in face of large-scale disasters. Several challenges must be addressed beforehand. While from a technical perspective the open questions we have identified yield substantial issues to be solved there are no fundamental show-stopper to allow the IoT to mitigate the impact of a disaster on network connectivity. The main question is thus not whether the IoT can be leveraged to improve disaster resilience, but rather to which extent and how it should be adopted.

INTRODUCTION TO COGNITIVE IOT

The Internet of Things (IoT) is a technological revolution that is bringing us into a new ubiquitous connectivity, computing, and communication era. The development of IoT depends on dynamic technical innovations in a number of fields. However, we argue that without comprehensive cognitive capability, IoT is just like an awkward stegosaurus: all brawn, no brains. To fulfill its potential and deal with growing challenges, we must take the cognitive capability into consideration and empower IoT with high-level intelligence. This leads to the development of an enhanced IoT paradigm called Brain-Empowered Internet of Things or Cognitive Internet of Things (CIoT), and investigate the involved key enabling techniques. Before gonging deep into the new concept CIoT and its enabling techniques, let's first share two interesting application scenarios that will probably come into our daily life in future:

Application Scenario 1: Let's imagine that it's Friday, after five days' hard work, I'd like to relax myself and watch a TV Soap Opera tonight. When time goes to the midnight, I become more and more sleepy and finally fall asleep on my sofa. Generally, I will wake up late on Saturday and feel very tired since I do not sleep well with the TV noise, the uncomfortable sofa and the fluctuating temperature all night long. Consequently, I have a dream that one day the TV, the sofa, and the air conditioner in my room could individually or cooperatively sense my movement, gesture, and/or voice, based on which they analyze my state (e.g., 'sleepy' or 'not sleepy'), and make corresponding decisions by themselves to comfort me, e.g., if I am in the state of 'sleepy', the TV itself gradually lowers or even turns off the voice, the sofa slowly changes itself to a bed, and the air conditioner dynamically adjusts the temperature suitable for sleep.

Application Scenario 2: Living in a modern city, traffic jams harass many of us. With potential traffic jams into consideration, every time when the source and the destination is clear, it is generally not easy for a driver to decide what the quickest route should be, especially when the driver is fresh to the city. Among many others, the following scheme may be welcome and useful for drivers: Suppose that there are cities of crowd sources, such as pre-deployed cameras, vehicles, drivers, and/or passengers, intermittently observe the traffic flow nearby and contribute their observations to a data center. The data center effectively fuses the crowd sourced observations to generate real-time traffic situation map and/or statistical traffic database. Then, every time when a driver tells his/her car the destination, the car will automatically query the data center, deeply analyze the accessed traffic situation information from the data center and meanwhile other cars/drivers' potential decisions, and intelligently selects the quickest route or a few top quickest routes for its driver.

IoT without a 'brain' is not enough to bring the expected convenient and comfortable life to us. These observations motivate us to develop the new paradigm Cognitive Internet of Things (CIoT). Cognitive Internet of Things (CIoT) is a new network paradigm, where (physical/virtual) things or objects are interconnected and behave as agents, with minimum human intervention, the things interact with each other following a context-aware perception-action cycle, use the methodology of understanding-by-building to learn from both the physical environment and social networks, store the learned semantic and/or knowledge in kinds of databases, and adapt themselves to changes or uncertainties via resource-efficient decision making mechanisms, with two primary objectives in mind:

Bridging the physical world (with objects, resources, etc) and the social world (with human demand, social behavior behavior, etc), together with themselves to form an intelligent physical-cyber-social (iPCS) system.

Enabling smart resource allocation, automatic network operation, and intelligent service provisioning.

System Architecture of CIOT

Our researches build on the network topology of CIoT whose sketch map is shown in Fig. 1. It includes core network and various access network domains. The core network is mainly made up of access router, wireless router, transmission router, etc. The access network domains include cognitive nodes, simple nodes and various terminals. The meanings of some components are illustrated as follows.

- **Autonomous Domain (AD):** It is an access network domain with autonomy and one of the following features. A high coupled and relative independent domain; A domain with distinct geographical feature; a network for organization, company, enterprise, etc; specially, autonomic devices in core network. If necessary, an AD can be divided into several Sub-ADs.
- **Cognitive Node:** It also called Cognitive Element (CE), refers a node which has the ability to autonomously optimize network performance according to current conditions.
- **Simple Node (SN):** It refers to a node without intelligence, which is relative to the cognitive node. There are different numbers of CEs in different ADs, maybe only one under the special circumstances. If there are multi CEs in an AD, two or more CEs can cooperate according to requirements.
- **Multi-Domain Cooperation (MDC):** For an application oriented to far broader network environment, the cooperative process of two or more ADs is called MDC.
- **Cognitive Agent (CA):** For a MDC, it refers the specific CEs selected from each domain to carry out cooperative assignments. There is different number of CAs in different domains, maybe only one.
- **Neighbor:** Two ADs with directly cooperative relationship are reciprocally called neighbors, and two ADs with cooperative relationship in virtue of other ADs are reciprocally called extended neighbors.

In CIoT, without artificial interventions, ADs divided, CAs selected and multi-domain cooperated are implemented autonomously. Some models based on Figure 1 are given as follows.

Framework of CIOT

Figure 2 presents a framework of CIoT.

Generally, CIoT serves as a transparent bridge between physical world (with general physical/virtual things, objects, resources, etc.) and social world (with human demand, social behavior, etc.), together with itself form an intelligent physical-

Figure 1. Autonomous Domain Models. This figure demonstrates the implementation of divided ADs and multi-domain.

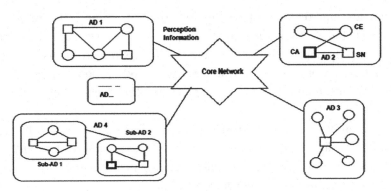

Figure 2. Framework of CIoT and fundamental cognitive tasks. This figure describes about framework of cognitive IoT along with its fundamental tasks.

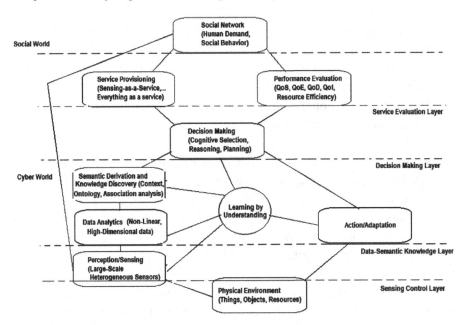

cyber-social (iPCS) system. From a bottom-up view, the cognitive process of the iPCS system consists of four major layers:

Sensing control layer has direct interfaces with physical environment, in which the preceptors sense the environment by processing the incoming stimuli and feedbacks observations to the upper layer, and the actuators act so as to control the preceptors via the environment.

Data-semantic-knowledge layer effectively analyzes the sensing data to form useful semantic and knowledge.

Decision-making layer uses the semantic and knowledge abstracted from the lower layer to enable multiple or even massive interactive agents to reason, plan and select the most suitable action, with dual functions to support services for human/social networks and stimulate action/adaption to physical environment.

Service evaluation layer shares important interfaces with social networks, in which on-demand service provisioning is provided to social networks, and novel performance metrics are designed to evaluate the provisioned services and feedback the evaluation result to the cognition process.

With a synthetic methodology learning-by-understanding located at the heart, the framework of CIoT includes five fundamental cognitive tasks, sequentially, Perception-action cycle, Massive data analytics, Semantic derivation and knowledge discovery, Intelligent decision-making, and On-demand service provisioning. Briefly, perception-action cycle is the most primitive cognitive task in CIoT with perception as the input from the physical environment and action as the output to it. On the other hand, on-demand service provisioning directly supports various services (e.g., Infrastructures- a-Service (IaaS), Platform-as-a-Service (PaaS), Sensing-as a-Service (SaaS), and more broadly Everything-as-a-service (EaaS) to human/social networks, which has been investigated recently.

INTELLIGENT DECISION MAKING FOR COGNITIVE INTERNET OF THINGS

Generally, decision-making in CIoT includes reasoning, planning and selecting. For reasoning and planning, the key concerns are analyzing the collected data and inferring useful information, which belong to data analysis in essence. To avoid illegibility, we refer to decision-making as selecting in this article. The task of selecting is common in CIoT, e.g., selecting the path in the smart traffic systems, choosing the channels for wireless transmission, and selecting the optimal service when there are multiple services simultaneously available. To summarize, selecting can be defined as the process of choosing an action from the action set. Motivated by the learning ability in cognitive radio networks, we study cognitive selecting in CIoT, which is characterized by having the ability to intelligently adjust the selecting based on the history information. Methodically, three kinds of cognitive selecting have been studied in the literature: Markovian decision process, multi-bandit armed problem and multi-agent learning. In comparison, the first two kinds are mainly for single decision maker while the third one is for multiple decision-makers. Since it is expected that there are a large number of decision makers (human or machine) in

CIoT, we focus on multi-agent learning. Since the selections of the decision-makers are interactive, we can formulate the multiple decision-making system as a game and then study multi-agent learning approaches. Specifically, we establish a framework for intelligent decision-making in CIoT, study intelligent decision-making in large-scale CIoT, and investigate the learning approaches with uncertain, dynamic and incomplete information.

We establish a framework for intelligent decision-making in CIoT, which is shown in Figure 3. Each decision-maker has semantic information and/or knowledge from the environment. Note that semantic information is generally generated by semantic derivation, while knowledge can be obtained from knowledge discovery or be given from social world in advance. In addition, it may have information about other decision makers if information exchange is available. However, if the information exchange is resource-consumed or even not available in some scenarios, a decision-maker does not have information about others. Using the information from the environment and others and taking into account its service demand, a decision-maker performs the cognitive selecting and outputs the decision result.

A COGNITIVE APPROACH TO IOT SECURITY

(Arbia 2014) et al. The systemic and cognitive approach for IoT security, illustrated in Figure 4 is made up of four nodes, namely person, people, technology, and intelligent object. To guarantee conformity in conception and implementation of secure applications, all these nodes must cooperate. The inclusion of the intelligent

Figure 3. The framework of intelligent decision-making in CIoT. This figure demonstrates about intelligent decision making framework.

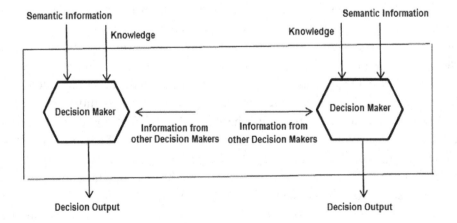

Figure 4. The proposed approach and its main elements. This figure explains about the primary elements involved in the proposed approach.

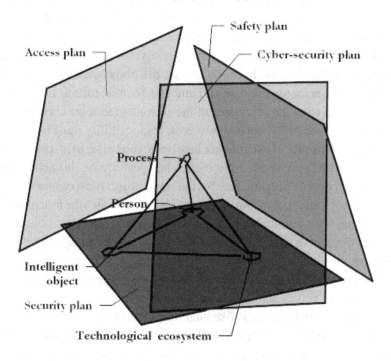

objects into such a complex system is a delicate issue for many reasons. First, this inclusion increases the complexity of the control process considerably. Second, the interaction between objects and people is difficult to address due to the increasing number of connected objects per person, the different levels of data sensitivity and security requirements. Finally, omnipresent objects lead modern technology to new applications and new services. Consequently, the resulting computing environment may involve humans, computers, sensors, RFID tags, network equipment and protocols, system software, and applications.

The connections among the nodes are dynamic and complex because they follow the environment characteristics. We refer to these connections as tensions. Tensions emphasize their roles of cooperation/conflict between the nodes. By modeling each tension and once the main actors of a security issues are identified, it will be easier to define the adequate solution by using our approach. Special interest must be given to understand these tensions and their security requirements. We name seven tensions between nodes: identification, trust, privacy, reliability, responsibility, safety and self-immunity.

To explain the difference between node and tension, we present an example of IoT application in a smart environment to ensuring comfortable services. We consider a scenario that involves a home owner, who plays the role of people, sensors and actuators within the house perform the role of intelligent objects, communication means and protocols depict the technological ecosystem and remote monitoring of heater represents the process. In this scenario, the home owner needs to identify the right sensor or actuator to adapt the ambient temperature to her preferences. The actuator has to trust the originator of the command to react correctly. Also, in order to avoid irresponsible people's mistakes, responsibility must be attributed to the right people. This process should not involve anyone else, then, privacy must be guaranteed. Safety of people and equipment when performing this action must be a priority to protect people's health. Finally, the smart object must ensure its immunity against physical or logical intrusion. In Figure 5 we present also the four planes within which the interactions among the nodes take place through the tensions, where we give a 2D perspective of each group of nodes. These planes are specified according to the relationships among the different triads of nodes. The Safety plane concerns process, person and technological ecosystem and involves the following tensions: privacy, safety and reliability. In this plane, the technological choice made by a person to perform a given process like analyzing, storing or distributing data must be done in a safe and reliable manner. The use of this approach during operations as process design, process change, operation and maintenance practices, incident reaction/response planes, etc. must respect privacy constraints in the overall IoT environment. The Security plane includes person, technological ecosystem and intelligent object and details related tensions namely: trust, privacy and identification. In the IoT, all kinds of objects and equipment are connected together through different technologies and networks. Then, users can profit to develop and benefit from new services and applications. However, it is imperative to identify various intervening entities rigorously, meet privacy requirements of users and data, and establish robust mechanisms of trust management to avoid access right violation and other privilege-related attacks. The Access plane contains process, person and intelligent object and implies their connected tensions: identification, safety and responsibility. The intelligent objects are able to interact with other networked entities (objects and/or persons) and store information related to a specific process. This interaction must be developed in a fluent manner that (i) identifies correctly the intelligent objects, (ii) respects safety rules of humans and equipment and (iii) precises convenient access rules and responsibilities for each entity. This is of great importance especially in ubiquitous environment where the presence of objects and humans can not be controlled. The Cyber-security plane includes process, technological ecosystem and intelligent object. The tensions considered for this plane are responsibility, trust and reliability. The objective is to produce an effort to ensure security properties of the IoT cyber environment against security risks. For example, testing the process operation and technology necessary to deploy security procedures during intelligent objects interaction is a serious task. The reliability of the equipment and communication means must be guaranteed, trust management techniques between intelligent objects must be implemented and responsibility states must be attributed.

Figure 5 depicts about the projections of 3D-pyramid under various planes. The main features of the actors (Riahi et al.) [38]involved in our model, namely: person, process, technological ecosystem and intelligent object, are highlighted and the role of each actor in our vision is presented

Person

Security concerns are always depending on people's interest and intentional/ unintentional behavior. The human resource is always involved in all the processes as a cause and/or an effect. Then, it is considered as the most basic node in the model.

Figure 5. Projections of the 3D-pyramid on each of the planes

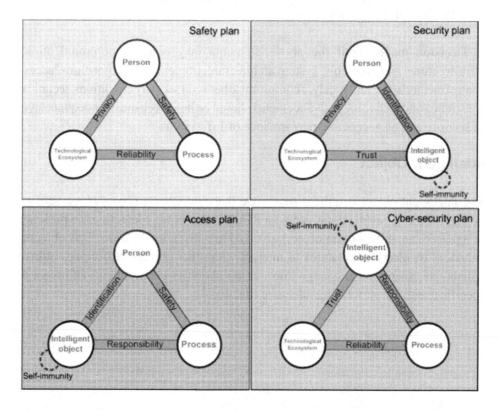

However, it is very difficult to control and to deal with people having different expectations, behaviors and technical knowledge.

Process

The process node is about a mean or a way to perform tasks in the IoT environment according to specific security conditions. Process must be in accordance with effective security policies to guarantee a sufficient level of security at different IoT architecture layers.

Technological Ecosystem

The third node is about the technological alternatives taken to guarantee acceptable IoT security level. There are five categories of information security elements:

- Security Design and Configuration;
- Identification and Authorization;
- Enclave internal;
- Enclave boundary;
- Physical and environmental.

Decision made about the above elements may concern communications infrastructures and protocols, system architecture, implemented algorithms, access control methods, etc. Obviously, a compromise between security conditions, technical constraints and technology advancements should be held to ensure an adequate level of security and an acceptable performance of IoT system.

Intelligent Object

The intelligent object is quite a new node that refers to an object like a sensing node (camera, X-ray machine, etc.), an RFID reader or tag (detecting the presence of a person, an animal or an object) involved in a given application. This object is enhanced by electronic features to interact with other objects. It becomes able to collaborate, share and exchange information about its environment, and react to specific events by performing adequate functioning.

In this Section we give a detailed definition of each tension and the related open research issues.

Privacy

Privacy represents the tension induced by the interaction between person and technological ecosystem. Data to be protected are necessarily related to human beings, thus their privacy is a mandatory objective of the IoT, due to the omnipresence of intelligent objects. Also, the misuse of technology is a cause of privacy violation.

Trust

Trust, the second tension that we consider in our model, links the intelligent object with the technological ecosystem. In the literature, many definitions of trust have been proposed. The first definition focuses on reliability trust and describes it as "the subjective probability by which an individual, A, expects that another individual, B, performs a given action on which its welfare depends. Here, we notice the presence of two main concepts: dependency and degree of trust (probability). The major drawback of this definition is the fact that trusting in a person is not enough to assume complete dependency on that person. The second definition is interested in decision trust and stipulates that "Trust is the extent to which someone is willing to depend on something or somebody in a given situation with a feeling of relative security, even though negative consequences are possible". Here four concepts are involved, namely: dependency, reliability, utility and risk. The third definition talks about trust management and précises it as "an approach to making decisions about interacting with something or someone we do not completely know, establishing whether we should proceed with the interaction or not".

Identification

The IoT envisions a huge number of devices (sensors, actuators, network equipment and so on), temporarily or permanently interconnected. In such conditions, identification and localization of a given object is a fundamental subject that concerns the general system operation including architecture, components, access rights, etc.

Reliability

According to ANSI, software reliability is defined as "the probability of failure-free software operation for a specified period of time in a specified environment" In the IoT context, this tension can be considered when handling unique and reliable entities addresses, managing data over the network or in case of effective use of device(s) for specific applications.

Safety

In real life, autonomous systems became more widespread. Their control software can be the cause of a random or unpredictable behavior. A similar situation must be controlled to avoid disastrous consequences for the whole system and the physical environment. Furthermore, sensors are widely used to feed databases with important information and signals. People may refuse participating in collective activities due to the privacy and safety concerns. Thus, safety seems to be a very important tension to avoid unexpected problems.

Responsibility

Responsibility is closely related to access rights or authorization privileges. For example, if a given IoT object is configured by one entity, it must be able to handle connections from other objects and distinguish their different access rights. Then, it grants authorizations according to these devices' access rights.

Self-Immunity

Frequently, nodes are used in distant and/or hostile areas. They became unprotected and exposed to physical attacks due to the site constraints such as unreliability of available wireless communication links, resource limitations, insufficient physical protection of nodes, absence of a robust trust management system, etc. The potential risks can be related to privacy (inventorying or rogue scanning, backward/forward tracing or tracking) or to security.

CITY MONITORING USING COGNITIVE IOT INTERFERENCE CONTROL

In the Urban Internet of Things (IoT), devices and systems are interconnected at the city scale to provide innovative services to the citizens. However, the traffic generated by the sensing and processing systems may overload local access networks. A coexistence problem arises where concurrent applications mutually interfere and compete for available resources. This effect is further aggravated by the multiple scales involved and heterogeneity of the networks supporting the urban IoT. One of the main contributions of this paper is the introduction of the notion of content-oriented cognitive interference control in heterogeneous local access networks supporting computing and data processing in the urban IoT. Here a network scenario where multiple communication technologies, such as Device-to-Device and Long Term Evolution (LTE), is considered. The focus is on city monitoring applications, where a video data stream generated by a camera system is remotely processed to detect objects. The cognitive network paradigm is extended to dynamically shape the interference pattern generated by concurrent data streams and induce a packet loss trajectory compatible with video processing algorithms. Numerical results show that the proposed cognitive transmission strategy enables a significant throughput increase of interfering applications for target accuracy of the monitoring application.

Application Scenario: Surveillance Camera and City Monitoring

In context of urban IoT, city monitoring and public safety applications become integral part of services enabled by the interconnection of multiple systems. These systems are inherently distributed systems, where local and edge processing are integrated to increase performance while reducing network traffic load. Herein, we consider a scenario where video stream from a surveillance camera is processed by edge-computing and sent to data management system. This can be extended in future to smart camera systems, where the cameras locally coordinate to determine encoding and transmission parameters of the data stream uploaded to the processing centers.

Video Streaming

Video is a sequence of images produced at a given rate. Unlike images, which only have a spatial component, a video has a temporal component as well.

Spatial compression: The individual pictures are broken into "macroblocks" which are transformed by Discrete Cosine Transform (DCT) from space domain to frequency domain. The basic MPEG-4 use 8 8 DCT where as H.264/MPEG-4 AVC (Advanced Video Coding) standard uses a 4 4 DCT-like integer transform. The transformed data, then, is quantized and encoded by entropy coding.

Temporal Compression

Temporal compression exploits the significant similarities that may interest pictures within the video stream captured at close time instant. This gives the opportunity to encode the video in less number of key (reference) frames and more number of compressed predicted frames following the reference frame. The decoder uses the preceding reference frame information to decode each of the predicted frames.

When an encoded frame is damaged, due to spatial compression, it affects the transform coefficients which lead to the multifold corruption in the decoded image content and resolution. The spatial propagation of errors may create artifacts that are detected as objects, or impair the ability of the algorithm to detect existing objects. On the other hand, due to temporal compression, the effect of the corruption varies depending on whether the damaged frame is a reference frame or not. If a reference frame is corrupted, the effect propagates through the entire Group of Pictures (GoP). Alternately, if a predicted frame is damaged, the effect is not so severe compare to losing a reference frame, as it may lose some information regarding motion vectors but key features in following frames are reconstructable. Here we consider H.264/AVC encoding which uses a Group of Pictures (GOP) for predictive coding of frames. The GOP starts with intra-coded frame (I-frame) followed by a number of inter coded frames e.g., Predicted frames (P-frames), Bi-directional predicted frames (B-frames) and can be of constant size or variable size.

The video streaming is done by packetizing the video data in a transport stream packet of fixed size 188 bytes. This data is then encapsulated by lower level protocols and sent over the network. The most popular video streaming protocols are HTTP Live Streaming (HLS) and Dynamic Adaptive Stream-ing over HTTP (DASH). If a packet is lost, the underlying TCP retransmits the packet. However, this setting may result in excessive delay in wireless environments, where channel variations and uncertainty induce non-negligible packet loss. On the other hand, protocols not providing packet retransmission such as Real time protocol (RTP) and UDP, packet loss are not recovered. In our proposed scheme, we reduce the effect of packet loss by cognitive interference control without using any retransmission overhead.

Video Processing

We use Object detection as a metric for measuring the quality of the video. Object detection has been one of the primary area of research in computer vision. Most of the object detection algorithms extract features from the image frames and employ a matching or detection algorithm with respect to reference image reported. Popular object detection algorithms include feature-based object detection, Viola-Jones object detection, object detection based on SVM (Support Vector Machines) and HOG (Histogram Oriented Gradient) features, Image segmentation and blob analysis. In this section, for illustrative purposes we used the Robust Features (SURF) based object recognition technique to recognize the point objects in each frame of a video with respect to a reference image.

Cognitive Interference Strategy

We take the approach of controlling the D2D transmission probability with respect to a given D2D transmission power in order to control the interference on the video transmission over LTE, achieving tolerable packet loss. We focus on topologies where the transmission power of the LTE user (transmitting power of the LTE user (transmitting the video stream) is sufficient to guarantee low packet failure probability in the absence of interference. A novel cognitive transmission technique is presented, where the transmission power and channel access scheme of the D2D link are based on the structure of the video data stream. In particular, we define a strategy that differentiates between groups of packets in the LTE video stream to improve the D2D throughput while limiting its impact on the ability of the monitoring and object detection algorithms. Our transmission strategy is based on the observation that damage to groups of LTE packets transporting a reference frame impacts the entire GOP. Transmission activity by the D2D link within a temporal period where packets associated with reference frames are transmitted generates a larger effective interference to the video quality with respect to the same activity performed during differential frames transmission. Increasingly, the effective interference generated by damage of reference frame is further enhanced by the error patterns induced to the video stream, which may result in the detection of non-existing objects.

The proposed technique only requires a statistical knowledge of the network channel coefficients, as opposed to the fine-granularity channel information required to control the SINR slot by slot. This a priori information is used by the D2D terminal to compute its transmission power, which determines the success probability at the D2D and LTE receivers conditioned on transmission. Then, the D2D terminal shapes its access strategy based on the video frame being transmitted by the LTE user. We call this as Frame Dependent Transmission Probability (FDTP). We compare FDTP scheme against a simple scheme where the D2D node determines transmission power and access probability to induce a fixed average failure probability to LTE packets. Then, in each slot the D2D terminal transmits with probability with a predetermined transmission power. We refer to this strategy as Fixed Probability (FP). Figure 6 shows a graphical representation and the finite state machines for the proposed schemes. In the following, we describe the proposed FDTP strategy.

Frame Dependent Transmission Probability (FDTP)

The D2D transmitter switches transmission access modes (LOW and HIGH) depending on whether the LTE is transmit-ting reference (I) or differential (B/P) frames. These modes correspond to transmission probabilities equal to I and D, respectively.

We implement this strategy in the LTE network scenario we described earlier. The LTE UE transmitting the video stream, sends a preamble message with the control information on the uplink to the eNode B about the video frame type (I, B/P) before sending new type of frame(s). As a GoP contains a series of frames and each frame contains hundreds of packets, the control message overhead for the preamble is much less compared to total number of packets. As we consider that the D2D devices are network assisted, they get downlink control information (DCI) about the preamble of the video data from the eNodeB. The D2D transmitter which is using the same spectrum as LTE then decides on its transmission probability based on the type of the frame mentioned in preamble. If the D2D decides not to transmit when I-packets (from reference frame) are being transmitted, it reduces

Figure 6. Finite State Machines for the interference control strategies

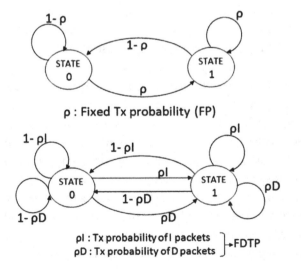

ρ : Fixed Tx probability (FP)

ρ_I : Tx probability of I packets
ρ_D : Tx probability of D packets } → FDTP

interference at the eNodeB for LTE uplink transmission. The eNodeB then reports this data-based CQI to the scheduler which then assigns higher MCS to the video transmitting UE until the I-frame packets are being transmitted, This way, it can prevent the I-frame loss. Contrary to this, when a differential frame (B/P) is being transmitted, D2D also transmits and interfere with LTE UE; as a result packet loss happens for differential frames. But since the reference frame is protected, this gives a gain in the received video quality compared to when the strategy is not applied and packet loss happens in reference frame as well.

Simulations and Numerical Results

We assess the performance of the proposed schemes by means of detailed network simulation and processing of video using NS-3 simulator [23] and MATLAB. First, we packetize the video with ffmpeg tool to generate transport stream which is inputted to the NS-3 simulator with a full end-to-end topology over the LTE protocol stack; then based on the interference scenario we generate video packet loss trace which we use in object detection by Computer Vision Toolbox in MATLAB. We used a H.264/AVC encoded video with high compression to evaluate the performance of video streaming over LTE in presence of D2D. The video is from a surveillance camera with fixed GoP size 128, where the background is almost fixed and only the foreground changes as objects move.

NS-3 version 3.23 with LTE protocol stack is used for simulation. We mimicked D2D communication using the same spectrum and radio channel as of LTE and chose the same bandwidth and central frequency of the channel and same set of

Table 3. LTE parameters for NS-3 simulation

LTE Parameters	Value
MAC Scheduler	Proportional Fair (PF)
RLC mode	RLC UM
eNodeB power	25 dB
UE Tx power	23 dB
D2D Tx Power	5 dB
Antenna Model	SISO
Path Loss model	Friis Propagation Loss
Uplink EARFCN	18100
UE-eNodeB distance	200 m
D2D distance	10 m

sub-channels for both LTE uplink and D2D communication. We consider the UE node is transmitting real-time video streaming data to a remote host via LTE and EPC network. The parameters used for experiments are mentioned in Table I. As we consider real-time delay sensitive application, we disable the HARQ and considered RLC UM (Unacknowledged Mode) in NS-3. In order to map the application packets with the PDUs at MAC layer, we disable the fragmentation and concatenation of the packets in LTE which reduces the overhead and complexity in computation, but does not violate the main idea of interference control. We used channel fading from the NS-3 trace file for Extended Pedestrian A model (EPA) as mentioned in 3GPP standard. Here, we experiment with both low speed (3 kmph) and high speed (10 kmph) fading for which we generate fading trace using MATLAB. For simulating the application, we considered a transport stream of the video with a stream of packets of equal size 188 bytes which we transmit via UDP transport protocol. We measure the object detection probability of received video with respect to the reference video (in absence of D2D interference) for variation of D2D throughput. Here we considered relative throughput of D2D with respect to reference throughput which assumes D2D always transmits and there is no packet loss. We performed the experiments in high speed fading and low speed fading scenarios. The transmission power of the UE and D2D is fixed throughout the experiment. A sample decoded received frame and its original version are shown in the Figure 7 to visualize the effect of interference on object detection.

We also define the efficiency of D2D operations with respect to the degradation caused to the video stream as follows:

$$\text{Efficiency} = \frac{\text{Object Detection Probability}}{1 - D2D \text{ Throughput}} \qquad (1)$$

Figure 7. Object Detection of original and received video

Figure 8. Contour plot for efficiency with respect to D2D transmission probability and D2D transmission power

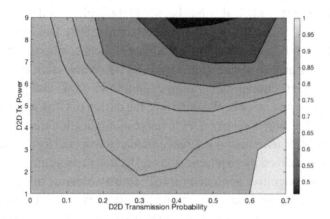

Figure 8 show the throughput of the D2D link and the efficiency as a function of the transmission probability and power. Intuitively, increasing the transmission power or the transmission probability increases the throughput. However, increasing the transmission power or the transmission probability also results in a larger degradation of the object detection probability. These two parameters can be jointly controlled to minimize the additional impact on the algorithm while increasing throughput. To achieve this, the D2D transmitter should change the parameters in the direction of throughput increase, while remaining in high efficiency regions. Preliminary results indicate that the two parameters can be chosen to generate minimum-impact failure patterns. A full study of this interesting effect, especially in relation with channel statistics is left to future studies. In this section a content-based cognitive trans-mission strategy for hybrid D2D/LTE networks supporting urban IoT applications was discussed. A monitoring application was considered, where video data streams are remotely processed to detect objects. The cognitive strategy shapes the transmission strategy to match the structure of video encoding and processing. Numerical results show significant throughput gain of the D2D link for a given performance degradation of the monitoring application with respect to the case when the interference process is fixed throughout the video transmission.

CONCLUSION

IoT is a novel concept that involves different technologies, human and non-human entities. Some recent efforts have been made in the direction of designing and deploying unifying architectures. However, none of these attempts specifically aimed at proposing a holistic vision of IoT security. IoT is a heterogeneous and mixed ubiquitous network, which has been widely applied in the field of modern intelligent service. However, the current IoT significantly lacks intelligence and cannot meet the application requirement. Also, practical solutions, if they exist, are intended to fulfill precise application needs (RFID, WSN, etc.). In this paper, we address this problem through integrating cognition into IoT, and present the CIoT and its architecture and models. We introduce the design of CIoT architecture. In the future, when attempting to realize a wide range of applications and services, it will become impossible to use a unique reference architecture model for real achievements. This openness and diversity require serious reflections concerning security issues. Our systemic and cognitive approach for IoT security remains still applicable even in the presence of the previously mentioned constraints and limitations of the IoT paradigm. In this study, we presented Internet of Things with architecture and design goals. We surveyed security and privacy concerns at different layers in IoTs. In addition, wc identified several open issues related to the security and privacy that need to be addressed by research community to make a secure and trusted platform for the delivery of future Internet of Things. We also discussed applications of IoT in real life scenarios. At the end a novel approach based on Cognitive IoT for city monitoring application was presented and detailed study was undertaken. In future, research on the IoTs will remain a hot issue.

REFERENCES

Atzori, L., Iera, A., & Morabito, G. (2010). The Internet of Things: A survey. *Computer Networks*, *54*(15), 2787–2805. doi:10.1016/j.comnet.2010.05.010

Atzori, L., Iera, A., & Morabito, G. (2011). SIoT: Giving a Social Structure to the Internet of Things. *IEEE Communications Letters*, *15*(11), 1193–1195. doi:10.1109/LCOMM.2011.090911.111340

Bansal, S. K. (2014). *Linux worm targets internet-enabled home appliances to mine cryptocurrencies*. Retrieved from http://thehackernews.com/2014/03/ linux-worm-targets-internet-enabled.html

Castellani, A. P., Dissegna, M., Bui, N., & Zorzi, M. (2012). WebIoT: A web application framework for the internet of things. *Proceedings of IEEE Wireless Communications and Networking Conference Workshops (WCNCW'12)*, 202-207.

Charilas, D. E., & Panagopoulos, A. D. (2010). A survey on game theory applications in wireless networks. *Computer Networks*, *54*(18), 3421–3430. doi:10.1016/j.comnet.2010.06.020

Dall'Anese, E., Kim, S., & Giannakis, G. B. (2011). Power Allocation for Cognitive Radio Networks under Channel Uncertainty. *Proc. of IEEE ICC.*

Ding, L., Shi, P., & Liu, B. (2010). The Clustering of Internet, Internet of Things and Social Network. *Proceedings of the 3rd International Symposium on Knowledge Acquisition and Modeling (KAM'10)*, 417-420.

Evans, D. (2011). *The internet of Things: How the next evolution of the Internet is changing everything.* Cisco Internet Business Solutions Group. Retrieved from http://www.cisco.com/web/about/ac79/docs/innov/IoT_IBSG_0411FINAL.pdf

Fortuna, C., & Mohorcic, M. (2009). Trends in the development of communication networks Cognitive networks. *Computer Networks*, *53*(9), 1354–1376. doi:10.1016/j.comnet.2009.01.002

Gates, B. (1995). *The Road Ahead.* New York: Penguin Books.

Guinard, D. (2009). Towards the web of things: Web mashups for embedded devices. *Proceedings of ACM MEM of WWW.*

Guinard, D., Trifa, V., Mattern, F., & Wilde, E. (2011). From the internet of things to the web of things: Resource-oriented architecture and best practice. In *Architecting the Internet of Things* (pp. 97–129). Springer Berlin. doi:10.1007/978-3-642-19157-2_5

Hong, S., Kim, D., Ha, M., Bae, S., Park, S. J., Jung, W., & Kim, J. (2010). SNAIL: An IP-based wireless sensor network approach to the internet of things. *IEEE Wireless Communications*, *17*(6), 34–42. doi:10.1109/MWC.2010.5675776

International Telecommunication Union (UIT). (2005). *ITU Internet Reports.* The Internet of Things.

Kamhoua, C. A., Pissinou, N., & Makki, S. K. (2010). Game Theoretic Analysis of Cooperation in Autonomous Multi Hop Networks. *Proc. of IEEE GLOBECOM.*

Kawsar, F., Fitton, D., & Sundramoorthy, V. (2010). Smart Objects as Building Blocks for the Internet of Things. *IEEE Internet Computing*, *14*(1), 44–51. doi:10.1109/MIC.2009.143

Li, X., Lu, R., Liang, X., Shen, X., Chen, J., & Lin, X. (2011). Smart community: an internet of things application. IEEE Communications Magazine, 49(11), 68-75.

Liu, W., Cui, Y., & Li, Y. (2015). Information systems security assessment based on system dynamics. *International Journal of Security and its Applications, 9*(2), 73-84.

Mitola, J., & Maguire, G. Q. (1999). Cognitive Radio: Making Software Radios More Personal. *IEEE Personal Communications, 6*(4), 13–18. doi:10.1109/98.788210

Ning, H., & Wang, Z. (2011). Future Internet of Things Architecture: Like Mankind Neural System or Social Organization Framework. *IEEE Communications Letters, 15*(4), 461–463. doi:10.1109/LCOMM.2011.022411.110120

Rabbachin, A., Quek, T. Q. S., & Shin, H. (2011). Cognitive Network Interference. *IEEE Journal on Selected Areas in Communications, 29*(2), 480–493. doi:10.1109/JSAC.2011.110219

Riahi, A., Natalizio, E., Challal, Y., Mitton, N., & Iera, A. (n.d.). *A systemic and cognitive approach for IoT security*. Retrieved from https://hal.inria.fr/hal-00863955

Rossi, B. (2015). *Gartner's Internet of Things predictions*. Information Age, Vitesse Media. Retrieved from http://www.information-age.com/technology/mobile-and-networking/123458905/gartners-internet-things-predictions

Senthil, S., & Senniappan, P. (2011). A Detailed Study on Energy Efficient Techniques for Mobile Ad hoc Networks. *International Journal of Computer Science Issues, 8*(1), 383–387.

Srivastava, V., & Motani, M. (2005). Cross-Layer Design: A Survey and the Road Ahead. *IEEE Communications Magazine, 43*(12), 112–119. doi:10.1109/MCOM.2005.1561928

Sundmaeker, H., Guillemin, P., Friess, P., & Woelfflé, S. (2010). Vision and Challenges for Realising the Internet of Things. *CERPIoT- Cluster of European Research Projects on the Internet of Things.*

Thomas, R. W., Friend, D. H., DaSilva, L. A., & Mackenzie, A. (2006). Cognitive Networks: Adaptation and Learning to Achieve End-to-End Performance Objectives. *IEEE Communications Magazine, 44*(12), 51–57. doi:10.1109/MCOM.2006.273099

Tong, X., & Ban, X. (2014). A hierarchical information system risk evaluation method based on asset dependence chain. *International Journal of Security and its Applications, 8*(6), 81-88.

Urgaonkar, R., & Neely, M. J. (2008). *Opportunistic Scheduling with Reliability Guarantees in Cognitive Radio Networks. Proc. of IEEE INFOCOM.*

Wang, Z., Ji, M., & Sadjadpour, H. R. (2009). *Cooperation-Multiuser Diversity Tradeoff in Wireless Cellular Networks. Proc. of IEEE GLOBECOM.*

Wright, A. (2011). Hacking cars. *Communications of the ACM, 54*(11), 18–19. doi:10.1145/2018396.2018403

Wu, M., Lu, T., Ling, F., Sun, J., & Du, H. (2010). Research on the architecture of Internet of things. *Proceedings of the 3rd International Conference on Advanced Computer Theory and Engineering (ICACTE'10)*, 484-487.

Yoshino, M., Kubo, H., & Shinkuma, R. (2009). Modeling User Cooperation Problem in Mobile Overlay Multicast as a Multiagent System. *Proc. of IEEE GLOBECOM.*

Zhang, H., & Zhu, L. (2011). Internet of Things: Key technology, Architecture and Challenging Problems. *Proceedings of IEEE International Conference on Computer Science and Automation Engineering (CSAE'11)*, 507-512.

Zhang, M., Wu, Q., Zheng, R., Wei, W., & Li, G. (2012). Research on Grade Optimization Self-Tuning Method for System Dependability Based on Autonomic Computing. *Journal of Computers, 7*(2), 333–340. doi:10.4304/jcp.7.2.333-340

Zhang, X., & Shin, K. G. (2010). DAC: Distributed Asynchronous Cooperation for Wireless Relay Networks. *Proc. of IEEE INFOCOM.*

Zheng, R., Wu, Q., Zhang, M., Li, G., Pu, J., & Wei, W. (2011). A self optimization mechanism of system service performance based on autonomic computing. *Journal of Computer Research and Development, 48*(9), 1676-1684. (in Chinese)

Zheng, R., Zhang, M., Wu, Q., Sun, S., & Pu, J. (2010). Analysis and application of bio-inspired multi-net security model. *International Journal of Information Security, 9*(1), 1–17. doi:10.100710207-009-0091-4

Compilation of References

Abdo, H., & Flaus, J. M. (2016). Uncertainty quantification in dynamic system risk assessment: A new approach with randomness and fuzzy theory. *International Journal of Production Research, 54*(19), 1–24. doi:10.1080/00207543.2016.1184348

Abraham, A., & Baikunth, N. (2000). *Hybrid Intelligent Systems: A Review of a decade of Research. School of Computing and Information Technology.* Autsralia *Technical Report Series (World Health Organization), 5/2000,* 1–55.

Abreu, F. B., Goulao, M., & Estevers, R. (1995). Toward the Design Quality Evaluation of OO Software Systems. *Proc. Fifth Int'l Conf. Software Quality.*

Abreu, F. B., & Melo, W. (1996). Evaluating the Impact of OO Design on Software Quality. *Proc. Third Int'l Software Metrics Symp.* 10.1109/METRIC.1996.492446

Ackoff, R. L., & Vergara, E. (1981). Creativity in problem solving and planning: A review. *European Journal of Operational Research, 7*(1), 1–13. doi:10.1016/0377-2217(81)90044-8

Adeli, A., & Neshat, M. (2010, March). A fuzzy expert system for heart disease diagnosis. In *Proceedings of International Multi Conference of Engineers and Computer Scientists (Vol. 1).* Academic Press.

Al Rahhal, M. M., Bazi, Y., AlHichri, H., Alajlan, N., Melgani, F., & Yager, R. R. (2016). Deep learning approach for active classification of electrocardiogram signals. *Information Sciences, 345,* 340–354. doi:10.1016/j.ins.2016.01.082

Albrecht, A. J., John, E., & Gaffney, J. R. (1983). Software Function, Source Lines of Code, and Development Effort Prediction: A Software Science Validation. *IEEE Transactions on Software Engineering, SE-9*(6), 639–648. doi:10.1109/TSE.1983.235271

Alexander, C. (2009). *Market risk analysis, value at risk models* (Vol. 4). John Wiley & Sons.

Al-Qutaish, R. E. (2010). Quality Models in Software Engineering Literature: An Analytical and Comparative Study. *The Journal of American Science*, 6(3), 166–175.

Al-Sakran, H. O. (2015). Framework architecture for improving healthcare information systems using agent technology. *International Journal of Managing Information Technology*, 7(1), 17–31. doi:10.5121/ijmit.2015.7102

AlSharqi, K., Abdelbari, A., Abou-Elnour, A., & Tarique, M. (2014). Zigbee Based Wearable Remote Healthcare Monitoring System for Elderly Patients. *International Journal of Wireless & Mobile Networks*, 6(3), 53–67. doi:10.5121/ijwmn.2014.6304

Altman, E. I., & Saunders, A. (1997). Credit risk measurement: Developments over the last 20 years. *Journal of Banking & Finance*, 21(11), 1721–1742. doi:10.1016/S0378-4266(97)00036-8

Alvarez, D. A. (2006). On the calculation of the bounds of probability of events using infinite random sets. *International Journal of Approximate Reasoning*, 43(3), 241–267. doi:10.1016/j.ijar.2006.04.005

Amara, D., & Rabai, L. B. A. (2017). Towards a new framework of software reliability measurement based on software metrics. *Procedia Computer Science*, 109, 725–730. doi:10.1016/j.procs.2017.05.428

Anderberg, M. R. (1773). Cluster analysis for applications. Academic Press.

Anoop, M. B., Balaji, R. K., & Lakshmanan, N. (2008). Safety assessment of austenitic steel nuclear power plant pipelines against stress corrosion cracking in the presence of hybrid uncertainties. *International Journal of Pressure Vessels and Piping*, 85(4), 238–247. doi:10.1016/j.ijpvp.2007.09.001

Anwar, M., Qadri, S. Q., & Sattar, A. R. (2013). Green Computing and Energy Consumption Issues in the Modern Age. *IOSR Journal of Computer Engineering*, 12(6), 91–98.

Architecture Journal. (n.d.). Retrieved from: www.architecturejournal.net

Arditti, F. D. (1967). Risk and the required return on equity. *The Journal of Finance*, 22(1), 19–36. doi:10.1111/j.1540-6261.1967.tb01651.x

Arunraj, N. S., Mandal, S., & Maiti, J. (2013). Modeling uncertainty in risk assessment: An integrated approach with fuzzy set theory and monte carlo simulation. *Accident; Analysis and Prevention*, 55, 242–255. doi:10.1016/j.aap.2013.03.007 PMID:23567215

Ashraf, M. A., Suzan, A., & Donald, C. W. (2007). Gaussian Versus Cauchy Membership Functions in Fuzzy PSO. *Proceedings of International Joint Conference on Neural Networks*.

Asiedu, E. (2002). On the determinants of foreign direct investment to developing countries: Is Africa different? *World Development*, *30*(1), 107–119. doi:10.1016/S0305-750X(01)00100-0

Atzori, L., Iera, A., & Morabito, G. (2010). The Internet of Things: A survey. *Computer Networks*, *54*(15), 2787–2805. doi:10.1016/j.comnet.2010.05.010

Atzori, L., Iera, A., & Morabito, G. (2011). SIoT: Giving a Social Structure to the Internet of Things. *IEEE Communications Letters*, *15*(11), 1193–1195. doi:10.1109/LCOMM.2011.090911.111340

Babuska, R. V., Vander, P. J., & Kaymak, U. (2002). Improved Covariance Estimation for Gustafson-Kessel Clustering. *Proceedings of the 2002 IEEE international conference on Fuzzy Systems*, 1081-1085.

Bache, K., & Lichman, M. (2013). *UCI machine learning repository*. University of California, School of Information and Computer Science. Retrieved from http://archive.ics.uci.edu/ml

Balke, W. T. (2002). A roadmap to personalized information systems by cognitive expansion of queries. In *Proc. of the EnCKompass Workshop on Content-Knowledge Management and Mass Personalization of E-Services (EnCKompass 2002)* (pp. 123-125). Academic Press.

Bansal, S. K. (2014). *Linux worm targets internet-enabled home appliances to mine cryptocurrencies*. Retrieved from http://thehackernews.com/2014/03/ linux-worm-targets-internet-enabled.html

Bansiya, J., & Davis, C. G. (2002). A hierarchical model for object-oriented design quality assessment. *IEEE Transactions on Software Engineering*, *28*(1), 4–17. doi:10.1109/32.979986

Baraldi, P., & Zio, E. (2008). A Combined Monte Carlo and Possibilistic Approach to Uncertainty Propagation in Event Tree Analysis. *Risk Analysis*, *28*(5), 1309–1326. doi:10.1111/j.1539-6924.2008.01085.x PMID:18631304

Bartram, S.M. (2015). *The Impact of Commodity Price Risk on Firm Value--An Empirical Analysis of Corporate Commodity Price Exposures*. Academic Press.

Basili, V. R., Briand, L. C., & Melo, W. L. (1996). A validation of object-oriented design metrics as quality indicators. *IEEE Transactions on Software Engineering*, *22*(10), 751–761. doi:10.1109/32.544352

Baudrit, C., & Dubois, D. (2006). Practical Representations of Incomplete Probabilistic Knowledge. *Computational Statistics & Data Analysis*, *51*(1), 86–108. doi:10.1016/j.csda.2006.02.009

Baudrit, C., Dubois, D., & Guyonnet, D. (2006). Joint Propagation and Exploitation of Probabilistic and Possibilistic Information in Risk Assessment. *IEEE Transactions on Fuzzy Systems*, *14*(5), 593–608. doi:10.1109/TFUZZ.2006.876720

Baudrit, C., Dubois, D., & Perrot, N. (2008). Representing parametric probabilistic models tainted with imprecision. *Fuzzy Sets and Systems*, *159*(15), 1913–1928. doi:10.1016/j.fss.2008.02.013

Bekaert, G., & Hodrick, R. J. (1993). On biases in the measurement of foreign exchange risk premiums. *Journal of International Money and Finance*, *12*(2), 115–138. doi:10.1016/0261-5606(93)90019-8

Berenji, H. R., & Khedkar, P. (1992). Learning and Tuning Fuzzy Logic Controllers through Reinforcements. *IEEE Transactions on Neural Networks*, *3*(5), 724–740. doi:10.1109/72.159061 PMID:18276471

Berleant, D. (1993). Automatically verified reasoning with both intervals and probability density functions. *Interval Computations*, *2*, 48–70.

Berleant, D., & Goodman-Strauss, C. (1998). Bounding the Results of Arithmetic Operations on Random Variables of Unknown Dependency Using Intervals. *Reliable Computing*, *4*(2), 147–165. doi:10.1023/A:1009933109326

Bevan, N. (1999). Quality in use: Meeting user needs for quality. *Journal of Systems and Software*, *49*(1), 89–96. doi:10.1016/S0164-1212(99)00070-9

Bezdek, J. C. (1983). *Pattern Recognition with Fuzzy Objective Function Algorithms*. New York: Plenum.

Bhatia, P. K., Choudhary, C., Mehtrotra, D., & Wahid, A. (2013). A review of the cognitive information retrieval concept, process and techniques. *Journal of Global Research in Computer Science*, *4*(3).

Bickmore, T. W., & Picard, R. W. (2004, April). Towards caring machines. In CHI'04 extended abstracts on Human factors in computing systems (pp. 1489-1492). ACM. doi:10.1145/985921.986097

Bickmore, T., Gruber, A., & Picard, R. (2005). Establishing the computer–patient working alliance in automated health behavior change interventions. *Patient Education and Counseling*, *59*(1), 21–30. doi:10.1016/j.pec.2004.09.008 PMID:16198215

Bielecki, T. R., & Rutkowski, M. (2013). *Credit risk: modeling, valuation and hedging*. Springer Science & Business Media.

Biswal, S. S., Mohanty, M. N., Das, S., & Sahu, B. (2016). Unconscious state analysis using intelligent signal processing techniques. *Advanced Science Letters*, *22*(2), 314–318. doi:10.1166/asl.2016.6856

Biswal, S., Das, A., Shalinee, S., & Mohanty, M. N. (2016). Design of telemedicine system for brain signal analysis. *International Journal of Telemedicine and Clinical Practices*, *1*(3), 250–264. doi:10.1504/IJTMCP.2016.077918

Bodie, Z., & Rosansky, V. I. (1980). Risk and return in commodity futures. *Financial Analysts Journal*, *36*(3), 27–39. doi:10.2469/faj.v36.n3.27

Boehm, B. W., Brown, J. R., Kaspar, J. R., Lipow, M., & MacCleod, G. (1978). *Characteristics of Software Quality*. Elsevier Science Ltd.

Bojadziev, G., & Bojadziev, M. (1995). *Fuzzy set, Fuzzy logic, application*. Singapore: World Scientific.

Bonabeau, E., Dorigo, M., & Theraulaz, G. (1999). *Swarm intelligence: from natural to artificial systems (No. 1)*. Oxford University Press.

Borlund, P. (2003). The concept of relevance in IR. *Journal of the Association for Information Science and Technology*, *54*(10), 913–925.

Bosch, N., D'Mello, S., Baker, R., Ocumpaugh, J., Shute, V., Ventura, M., & Zhao, W. (2015, March). Automatic detection of learning-centered affective states in the wild. In *Proceedings of the 20th international conference on intelligent user interfaces* (pp. 379-388). ACM.

Bouissou, O., Goubault, E., Larrecq, J. G., & Putot, S. (2011). A generalization of P-boxes to affine arithmetic. *Computing*. doi:10.100700607-011-0182-8

Braille, R. T., & Bollersley, T. (1990). A multivariate generalized ARCH approach to modeling risk premia in forward foreign exchange rate markets. *Journal of International Money and Finance*, *9*(3), 309–324. doi:10.1016/0261-5606(90)90012-O

Braun, A., Schmeiser, H., & Schreiber, F. (2017). Portfolio optimization under solvency II: Implicit constraints imposed by the market risk standard formula. *The Journal of Risk and Insurance*, *84*(1), 177–207. doi:10.1111/jori.12077

Brehme, U., Stollberg, U., Holz, R., & Schleusener, T. (2008). ALT pedometer—New sensor-aided measurement system for improvement in oestrus detection. *Computers and Electronics in Agriculture, 62*(1), 73–80. doi:10.1016/j.compag.2007.08.014

Briand, L. C., Morasca, S., & Basili, V. R. (1996). Property-based software engineering measurement. *IEEE Transactions on Software Engineering, 22*(1), 68–86. doi:10.1109/32.481535

Briand, L. C., Wüst, J., Daly, J. W., & Porter, D. V. (2000). Exploring the relationship between design measures and software quality in object-oriented systems. *Journal of Systems and Software, 51*(3), 245–273. doi:10.1016/S0164-1212(99)00102-8

Buckley, P. J., & Ghauri, P. N. (2015). *International business strategy: theory and practice*. Routledge. doi:10.4324/9781315848365

Bush, M. E., & Fenton, N. E. (1990). Software measurement: A conceptual framework. *Journal of Systems and Software, 12*(3), 223–231. doi:10.1016/0164-1212(90)90043-L

Castellani, A. P., Dissegna, M., Bui, N., & Zorzi, M. (2012). WebIoT: A web application framework for the internet of things. *Proceedings of IEEE Wireless Communications and Networking Conference Workshops (WCNCW'12)*, 202-207.

Charilas, D. E., & Panagopoulos, A. D. (2010). A survey on game theory applications in wireless networks. *Computer Networks, 54*(18), 3421–3430. doi:10.1016/j.comnet.2010.06.020

Chattopadhyay, S., Pratihar, D. K., & Sarkar, S. C. D. (2011). A Comparative Study of Fuzzy c-Means Algorithm and Entropy-Based Fuzzy Clustering Algorithms. *Computer Information, 30*, 701–720.

Chavez-Demoulin, V., Embrechts, P., & Neslehova, J. (2006). Quantitative models for operational risk: Extremes, dependence and aggregation. *Journal of Banking & Finance, 30*(10), 2635–2658.

Chawla, S. (2012). Review of MOOD and QMOOD metric sets. *International Journal of Advanced Research in Computer Science and Software Engineering, 3*(3), 448–451.

Chen, Z., Zhao, L., & Lee, K. (2010). Environmental risk assessment of offshore produced water discharges using a hybrid fuzzy-stochastic modeling approach. *Environmental Modelling & Software, 25*(6), 782–792. doi:10.1016/j.envsoft.2010.01.001

Chhikara, A., Chhillar, R.S., & Khatri, S. (2011). *Applying Object Oriented Metrics to C#(C Sharp) programs*. Academic Press.

Chi, C. L., Street, W. N., & Katz, D. A. (2010). A decision support system for cost-effective diagnosis. *Artificial Intelligence in Medicine*, *50*(3), 149–161. doi:10.1016/j.artmed.2010.08.001 PMID:20933375

Chidamber, S. R., & Kemerer, C. F. (1994). A metrics suite for object oriented design. *IEEE Transactions on Software Engineering*, *20*(6), 476–493. doi:10.1109/32.295895

Chistropher, H. F., & Bharvikar, R. (2002). Quantification of variability and Uncertainty: A case study of power Plant Hazardous Air Pollutant Emission. In D. Paustenbach (Ed.), *Human Ecological Risk Analysis* (pp. 587–617). New York: John Wiley and Sons.

Choudhary, S. (2014). A Survey on Green Computing Techniques. *International Journal of Computer Science and Information Technologies*, *5*(5), 6248–6252.

Chowdhury, C.R., Chatterjee, A., Sardar, A., Agarwal, S. & Nath, A. (2013). A Comprehensive study on Cloud Green Computing: To Reduce Carbon Footprints Using Clouds. *International Journal of Advanced Computer Research*, 1-9.

Clarke, A. C. (2016). *2001: a space odyssey*. Penguin.

Clifton Green, T., & Figlewski, S. (1999). Market risk and model risk for a financial institution writing options. *The Journal of Finance*, *54*(4), 1465–1499.

Codd, E. F. (1970). A relational model of data for large shared data banks. *Communications of the ACM*, *13*(6), 377–387. doi:10.1145/362384.362685

Connellan, K., Gaardboe, M., Riggs, D., Due, C., Reinschmidt, A., & Mustillo, L. (2013). Stressed spaces: Mental health and architecture. *HERD: Health Environments Research & Design Journal*, *6*(4), 127–168. doi:10.1177/193758671300600408 PMID:24089185

Constine, J. (2017). Facebook Rolls Out AI to Detect Suicidal Posts before They're Reported. *Tech Crunch*. Retrieved from https://techcrunch.com/2017/11/27/facebook-ai-suicide-prevention/

Conte, S. D., & De Boor, C. (1981). *Elementary Numerical Analysis: An Algorithmic Approach* (3rd ed.). McGraw-Hill.

Conte, S. D., Dunsmore, H. D., & Shen, V. Y. (1986). *Software Engineering Metrics and Models*. Menlo Park, CA: Benjamin-Cummings.

Cooke, B., Jiang, H., & Heerman, K. (2017). *Outlook for U.S. Agricultural Trade: August 2017*. United States Department of Agriculture Economic Research Service. Retrieved from https://www.ers.usda.gov/topics/international-markets-trade/us-agricultural-trade/outlook-for-us-agricultural-trade/

Craven, M. W., & Shavlik, J. W. (1997). Using neural networks for data mining. *Future Generation Computer Systems, 13*(2-3), 211–229. doi:10.1016/S0167-739X(97)00022-8

Crespo, L. G., Kenny, S. P., & Giesy, D. P. (2012). Reliability analysis of polynomial systems subject to P-box uncertainties. *Mechanical Systems and Signal Processing*. doi:10.1016/j.ymssp.2012.08.012

Crevier, D. (1993). *AI: The tumultuous history of the search for artificial intelligence*. Basic Books.

Cruz, M. G. (2002). *Modeling, measuring and hedging operational risk*. Chichester, UK: John Wiley & Sons.

Cummins, N., Hantke, S., Schnieder, S., Krajewski, J., & Schuller, B. (2017). Classifying the context and emotion of dog barks: a comparison of acoustic feature representations. *Compare: A Journal of Comparative Education, 40*(05), 37.

Czogala, E. & Leski, J. (2000). *Neuro-Fuzzy Intelligent Systems, Studies in Fuzziness and Soft Computing*. Springer Verlag.

Dall'Anese, E., Kim, S., & Giannakis, G. B. (2011). Power Allocation for Cognitive Radio Networks under Channel Uncertainty. *Proc. of IEEE ICC*.

Das, S.R., Mishra, D., & Rout, M. (2017). A hybridized ELM-Jaya forecasting model for currency exchange prediction. *Journal of King Saud University-Computer and Information Sciences*.

Das, R., Turkoglu, I., & Sengur, A. (2009). Effective diagnosis of heart disease through neural networks ensembles. *Expert Systems with Applications, 36*(4), 7675–7680. doi:10.1016/j.eswa.2008.09.013

Destercke, S., & Dubois, D. (2009). The role of generalised P-boxes in imprecise probability models. *6th International Symposium on Imprecise Probability: Theories and Applications*, Durham, UK.

Devi, C. S., Ramani, G. G., & Pandian, J. A. (2014). Intelligent E-healthcare management system in medicinal science. *International Journal of Pharm Tech Research, 6*(6), 1838–1845.

Dictionary, O. E. (2017). *Oxford English dictionary online*. Retrieved from https://en.oxforddictionaries.com/

Dillon, T., Wu, C., & Chang, E. (2010). Cloud Computing: Issues and Challenges. *2010 24th IEEE International Conference on Advanced Information Networking and Applications*, 27-33, DOI 10.1109/AINA.2010.187

Ding, L., Shi, P., & Liu, B. (2010). The Clustering of Internet, Internet of Things and Social Network. *Proceedings of the 3rd International Symposium on Knowledge Acquisition and Modeling (KAM'10)*, 417-420.

Domowitz, I., & Hakkio, C. S. (1985). Conditional variance and the risk premium in the foreign exchange market. *Journal of International Economics*, 19(1-2), 47–66. doi:10.1016/0022-1996(85)90018-2

Dong, X., Feng, S., & Sadka, R. (2017). Liquidity risk and mutual fund performance. *Management Science*, mnsc.2017.2851. doi:10.1287/mnsc.2017.2851

Dowd, K. (2007). *Measuring market risk*. John Wiley & Sons.

Dowling, G. R., & Staelin, R. (1994). A model of perceived risk and intended risk-handling activity. *The Journal of Consumer Research*, 21(1), 119–134. doi:10.1086/209386

Dumas, B., & Solnik, B. (1995). The world price of foreign exchange risk. *The Journal of Finance*, 50(2), 445–479. doi:10.1111/j.1540-6261.1995.tb04791.x

Dunn, J. C. (1773). A Fuzzy Relative of the ISODATA Process and its Use in Detecting Compact Well-Separated Clusters. *Journal of Cybernetics,Vol.*, 3(3), 32–57. doi:10.1080/01969727308546046

Dutta, P. (2013). An approach to deal with aleatory and epistemic uncertainty within the same framework: Case study in risk assessment. *International Journal of Computers and Applications*, 80(12).

Dutta, P. (2015). Uncertainty modeling in risk assessment based on Dempster–Shafer theory of evidence with generalized fuzzy focal elements. *Fuzzy Information and Engineering*, 7(1), 15–30. doi:10.1016/j.fiae.2015.03.002

Dutta, P. (2017). Modeling of variability and uncertainty in human health risk assessment. *MethodsX*, 4, 76–85. doi:10.1016/j.mex.2017.01.005 PMID:28239562

Dutta, P. (2018). Human Health Risk Assessment Under Uncertain Environment and Its SWOT Analysis. *The Open Public Health Journal*, 11(1), 72–92. doi:10.2174/1874944501811010072

Dutta, P., & Ali, T. (2011). A Proposal for Construction of P-box. *International Journal of Advanced Research in Computer Science*, 2(5), 221–225.

Dutta, P., & Ali, T. (2015). Aleatory and Epistemic Uncertainty Quantification. In *Applied Mathematics* (pp. 209–217). New Delhi: Springer.

Dutta, P., & Hazarika, G. C. (2017). Construction of families of probability boxes and corresponding membership functions at different fractiles. *Expert Systems: International Journal of Knowledge Engineering and Neural Networks*, 34(3), e12202. doi:10.1111/exsy.12202

Eisenberg, M. B. (1988). Measuring relevance judgments. *Information Processing & Management*, 24(4), 373–389. doi:10.1016/0306-4573(88)90042-8

Elavathingal, E. E., & Sethuramalingam, T. K. (n.d.). A Survey of EHealthcare Systems Using Wearable Devices and Medical Sensor Networks. *KJER, 1*(1).

EPA US. (2004). *Risk Assessment Guidance for Superfund, Volume I: Human Health Evaluation Manual (Part E, Supplemental Guidance for Dermal Risk Assessment)*. Office of Emergency and Remedial Response, EPA/540/R/99/005, Interim, Review Draft. United States Environmental Protection Agency.

Etzkorm, L., & Delugach, H. (2000). Semantic Metrics Suite for Object-Oriented Design. *34th International Conference on Technology of Object-Oriented Languages and Systems*.

Evans, D. (2011). *The internet of Things: How the next evolution of the Internet is changing everything*. Cisco Internet Business Solutions Group. Retrieved from http://www.cisco.com/web/about/ac79/docs/innov/IoT_IBSG_0411FINAL.pdf

Fakhfakh, R., Feki, G., Ammar, A. B., & Amar, C. B. (2016, October). Personalizing information retrieval: A new model for user preferences elicitation. In SMC (pp. 2091-2096). Academic Press.

Faloutsos, C. (1985). Access methods for text. *ACM Computing Surveys*, 17(1), 49–74. doi:10.1145/4078.4080

Farooq, S. U., Quadri, S. M. K., & Ahmad, N. (2012). Metrics, Models and Measurements in Software Reliability. *IEEE 10th International Symposium on Applied Machine Intelligence and Informatics (SAMI)*, 441 - 449.

Fenton, N., & Bieman, J. (2014). Software Metrics: A Rigorous and Practical Approach (3rd ed.). Academic Press. doi:10.1201/b17461

Fenton, N. (1994). Software Measurement: A Necessary Scientific Basis. *IEEE Transactions on Software Engineering*, 20(3), 199–206. doi:10.1109/32.268921

Fenton, N., & Pfleeger, S. L. (1996). *Software Metrics: A Rigorous & Practical Approach* (2nd ed.). International Thomson Computer Press.

Fernandez, P., Ortiz, A., & Acan, I. F. (2017). Market Risk Premium Used in 71 Countries in 2016: A Survey with 6,932 Answers. *Journal of International Business Research and Marketing, 2*(6), 23–31. doi:10.18775/jibrm.1849-8558.2015.26.3003

Fernandez, R., & Picard, R. W. (2003). Modeling drivers' speech under stress. *Speech Communication, 40*(1), 145–159. doi:10.1016/S0167-6393(02)00080-8

Ferson, S., Kreinovich, V., Ginzburg, L., Myers, D. S., & Sentz, K. (2003). *Construction of probability boxes and Dempster-Shafer structures*. Tech. Rep. SAND2002-4015, Sandia National Laboratories.

Ferson, S., & Donald, S. (1998). In A. Mosleh & R. A. Bari (Eds.), *Probability Bounds Analysis, Probabilistic Safety Assessment and Management* (pp. 1203–1208). New York, NY: Springer-Verlag.

Ferson, S., Nelsen, R., Hajagos, J., Berleant, D., Zhang, J., Tucker, W., ... Oberkampf, W. (2004). *Dependence in probabilistic modelling, Dempster–Shafer theory and probability bounds analysis. Tech. rep.* Sandia National Laboratories.

Figueiredo & Gomide. (1999). Design of Fuzzy Systems Using Neuro-Fuzzy Networks. *IEEE Transactions on Neural Networks, 10*(4), 815-827.

Flage, R., Baraldi, P., Zio, E., & Aven, T. (2010) Possibility-probability transformation in comparing different approaches to the treatment of epistemic uncertainties in a fault tree analysis. *Reliability, Risk and Safety - Proceedings of the ESREL, Conference*, 714-721.

Flage, R., Baraldi, P., Zio, E., & Aven, T. (2013). Probabilistic and Possibilistic treatment of epistemic uncertainties in fault tree analysis. *Risk Analysis, 33*, 121–133. doi:10.1111/j.1539-6924.2012.01873.x PMID:22831561

Fletcher, R. R., Poh, M. Z., & Eydgahi, H. (2010, August). Wearable sensors: opportunities and challenges for low-cost health care. In *Engineering in Medicine and Biology Society (EMBC), 2010 Annual International Conference of the IEEE* (pp. 1763-1766). IEEE.

Forkan, A., Khalil, I., & Tari, Z. (2013, September). Context-aware cardiac monitoring for early detection of heart diseases. In *Computing in Cardiology Conference (CinC), 2013* (pp. 277-280). IEEE.

Fortuna, C., & Mohorcic, M. (2009). Trends in the development of communication networks Cognitive networks. *Computer Networks, 53*(9), 1354–1376. doi:10.1016/j.comnet.2009.01.002

Fraley, C., & Raftery, A. (1999). MCLUST: Software for Model-Based Cluster Analysis. *J. Classificat., 16*(2), 297–306. doi:10.1007003579900058

Fraser, D., & Duncan, I. J. (1998). *'Pleasures', 'pains' and animal welfare: toward a natural history of affect.* Academic Press.

Frechet, M. (1935). Generalisations du theoreme des probabilite's totals. *Fundamenta Mathematicae, 25*, 379–387. doi:10.4064/fm-25-1-379-387

Gabriel, S. C., & Baker, C. B. (1980). Concepts of business and financial risk. *American Journal of Agricultural Economics, 62*(3), 560–564. doi:10.2307/1240215

Gallitschke, J., Seifried, S., & Seifried, F. T. (2017). Interbank interest rates: Funding liquidity risk and XIBOR basis spreads. *Journal of Banking & Finance, 78*, 142–152. doi:10.1016/j.jbankfin.2017.01.002

Garbarino, M., Lai, M., Bender, D., Picard, R. W., & Tognetti, S. (2014, November). Empatica E3—A wearable wireless multi-sensor device for real-time computerized biofeedback and data acquisition. In *Wireless Mobile Communication and Healthcare (Mobihealth), 2014 EAI 4th International Conference on* (pp. 39-42). IEEE.

Garg, S. K., & Buyya, R. (2012). *Green Cloud computing and Environmental Sustainability.* Cloud computing and Distributed Systems (CLOUDS) Laboratory. Retrieved from www.cloudbus.org/papers/Cloud-EnvSustainability2011.pdf

Garg, S. K., & Buyya, R. (n.d.). Green Cloud computing and Environmental Sustainability. Cloud computing and Distributed Systems (CLOUDS). *Laboratory.*

Gates, B. (1995). *The Road Ahead.* New York: Penguin Books.

Genther, H., Runkler, T. A., & Glesner, M. (1994). Defuzzification based on fuzzy clustering. *Proceeding of IEEE World Congress on Computational Intelligence.*

Gilb, T. (1988). *Principles of Software Engineering Management* (2nd ed.). Reading, MA: Addison-Wesley.

Godsk, T., & Kjærgaard, M. B. (2011, August). High classification rates for continuous cow activity recognition using low-cost GPS positioning sensors and standard machine learning techniques. In *Industrial Conference on Data Mining* (pp. 174-188). Springer.

Goel, A. L. (1985). Software reliability models: Assumptions, limitations and applicability. *IEEE Transactions on Software Engineering*, *11*(12), 1411–1423. doi:10.1109/TSE.1985.232177

Goertz, C. E., Shuert, C. R., Mellish, J. A. E., Skinner, J. P., Woodie, K., Walker, K. A., ... Berngartt, R. K. (2017). Best practice recommendations for the use of fully implanted telemetry devices in pinnipeds. *Animal Biotelemetry*, *5*(1), 13. doi:10.118640317-017-0128-9

Gorrab, A., Kboubi, F., & Ghezala, H. (2017). Social Information Retrieval and Recommendation: state-of-the-art and future research. *HAL*. Retrieved from: https://hal.archives-ouvertes.fr/hal-01444570/document

Gorrab, A., Kboubi, F., Ghezala, H. B., & Le Grand, B. (2016, November). Towards a dynamic and polarity-aware social user profile modeling. In *Computer Systems and Applications (AICCSA), 2016 IEEE/ACS 13th International Conference of* (pp. 1-7). IEEE.

Grady, R. B., & Caswell, D. (1987). *Software Metrics: Establishing a Company-Wide Program*. Englewood Cliffs, NJ: Prentice-Hall.

Granger, M. M., & Henrion, M. (1992). *Uncertainty- A guide to dealing uncertainty in Quantitative risk and policy analysis*. Cambridge University Press.

Griffin, N. L., & Lewis, F. D. (1989, October). A rule-based inference engine which is optimal and VLSI implementable. In *Tools for Artificial Intelligence, 1989. Architectures, Languages and Algorithms, IEEE International Workshop on* (pp. 246-251). IEEE.

Guay, W. R. (1999). The sensitivity of CEO wealth to equity risk: An analysis of the magnitude and determinants. *Journal of Financial Economics*, *53*(1), 43–71. doi:10.1016/S0304-405X(99)00016-1

Guinard, D. (2009). Towards the web of things: Web mashups for embedded devices. *Proceedings of ACM MEM of WWW*.

Guinard, D., Trifa, V., Mattern, F., & Wilde, E. (2011). From the internet of things to the web of things: Resource-oriented architecture and best practice. In *Architecting the Internet of Things* (pp. 97–129). Springer Berlin. doi:10.1007/978-3-642-19157-2_5

Gustafson, D. E., & Kessel, W. C. (1978). Fuzzy Clustering with a Fuzzy Covariance Matrix. *IEEE Conference on Adaptive Processes*, *17*(1), 761-766.

Guyonnet, D., Bourgine, B., Dubois, D., Fargier, H., Côme, B., & Chilès, J. P. (2003). Hybrid approach for addressing uncertainty in risk assessments. *Journal of Environmental Engineering*, *126*, 68–78. doi:10.1061/(ASCE)0733-9372(2003)129:1(68)

Guyonnet, D., Côme, B., Perrochet, P., & Parriaux, A. (1999). Comparing two methods for addressing uncertainty in risk assessments. *Journal of Environmental Engineering*, *125*(7), 660–666. doi:10.1061/(ASCE)0733-9372(1999)125:7(660)

Halstead, M. H. (1977). *Elements of Software Science (Operating and programming systems series)*. New York: Elsevier Science Inc.

Han, J., & Kamber, M. (2001). *Data Mining*. San Francisco, CA: Morgan Kaufmann Publishers.

Hannan, S. A., Mane, A. V., Manza, R. R., & Ramteke, R. J. (2010, December). Prediction of heart disease medical prescription using radial basis function. In *Computational Intelligence and Computing Research (ICCIC), 2010 IEEE International Conference on* (pp. 1-6). IEEE.

Harrison, R., & Steve, J. (1998). An Evaluation of the MOOD Set of Object-Oriented Software Metrics. *IEEE Transactions on Software Engineering*, *24*(6), 491–496. doi:10.1109/32.689404

Hathaway, R., Bezdek, J., & Hu, Y. (2000). Generalized fuzzy -means clustering strategies using norm distances. *IEEE Transactions on Fuzzy Systems*, *8*(5), 576–582. doi:10.1109/91.873580

Haykin, S. (1999). Neural Network (2nd ed.). Academic Press.

Healey, J., & Picard, R. (1998, May). Digital processing of affective signals. In *Acoustics, Speech and Signal Processing, 1998*. Proceedings of the 1998 IEEE International Conference on (Vol. 6, pp. 3749-3752). IEEE.

Healey, J. A., & Picard, R. W. (2005). Detecting stress during real-world driving tasks using physiological sensors. *IEEE Transactions on Intelligent Transportation Systems*, *6*(2), 156–166. doi:10.1109/TITS.2005.848368

Helali, R. G. M. (2015). A Comparison between Semantic and Syntactic Software Metrics. *International Journal of Advanced Research in Computer and Communication Engineering*, *4*(2).

Helwatkar, A., Riordan, D., & Walsh, J. (2014, September). Sensor technology for animal health monitoring. *International Conference on Sensing Technology, ICST*.

Hendler, J. (2008). Avoiding another AI winter. *IEEE Intelligent Systems*, *23*(2), 2–4. doi:10.1109/MIS.2008.20

Hendricks, K. B., & Singhal, V. R. (2005). An empirical analysis of the effect of supply chain disruptions on long-run stock price performance and equity risk of the firm. *Production and Operations Management*, *14*(1), 35–52. doi:10.1111/j.1937-5956.2005.tb00008.x

Hernandez Rivera, J. (2015). *Towards wearable stress measurement* (Doctoral dissertation). Massachusetts Institute of Technology.

Hernandez, J., McDuff, D., Fletcher, R., & Picard, R. W. (2013, March). Inside-out: Reflecting on your inner state. In *Pervasive Computing and Communications Workshops (PERCOM Workshops), 2013 IEEE International Conference on* (pp. 324-327). IEEE.

Hertzum, M., Søes, H., & Frøkjær, E. (1993). Information retrieval systems for professionals: A case study of computer supported legal research. *European Journal of Information Systems*, *2*(4), 296–303. doi:10.1057/ejis.1993.40

Hodrick, R. J., & Srivastava, S. (1984). An investigation of risk and return in forward foreign exchange. *Journal of International Money and Finance*, *3*(1), 5–29. doi:10.1016/0261-5606(84)90027-5

Hofstadter, D. (1980). *Gödel, Escher, Bach: An Eternal Golden Braid*. New York: Basic Books.

Hong, S., Kim, D., Ha, M., Bae, S., Park, S. J., Jung, W., & Kim, J. (2010). SNAIL: An IP-based wireless sensor network approach to the internet of things. *IEEE Wireless Communications*, *17*(6), 34–42. doi:10.1109/MWC.2010.5675776

Hoque, M. E. (2008, October). Analysis of speech properties of neurotypicals and individuals diagnosed with autism and down. In *Proceedings of the 10th international ACM SIGACCESS conference on Computers and accessibility* (pp. 311-312). ACM.

Howell, L. D., & Chaddick, B. (1994). Models of political risk for foreign investment and trade: An assessment of three approaches. *The Columbia Journal of World Business*, *29*(3), 70–91. doi:10.1016/0022-5428(94)90048-5

Hristidis, V., Koudas, N., & Papakonstantinou, Y. (2001). Prefer. *Proceedings of the 2001 ACM SIGMOD international conference on Management of data - SIGMOD 01*. 10.1145/375663.375690

Hubel, D. H., & Wiesel, T. N. (1964). Effects of monocular deprivation in kittens. *Naunyn-Schmiedeberg's Archives of Pharmacology, 248*(6), 492–497. doi:10.1007/BF00348878 PMID:14316385

Hudson, C. D., Bradley, A. J., Breen, J. E., & Green, M. J. (2012). Associations between udder health and reproductive performance in United Kingdom dairy cows. *Journal of Dairy Science, 95*(7), 3683–3697. doi:10.3168/jds.2011-4629 PMID:22720926

IEEE Standard 1061, Software Quality Metrics Methodology, IEEE Computer Society Press, New York, 2009.

IEEE Standard 610.12-1990, Glossary of Software Engineering Terminology, IEEE Computer Society Press, New York, 1990.

IEEE Std 1633 - IEEE Recommended Practice on Software Reliability, 2008.

Ingwersen, P., & Willett, P. (1995). An introduction to algorithmic and cognitive approaches for information retrieval. *Libri, 45*(3-4), 160–177. doi:10.1515/libr.1995.45.3-4.160

Innal, F., Chebila, M., & Dutuit, Y. (2016). Uncertainty handling in safety instrumented systems according to IEC 61508 and new proposal based on coupling Monte Carlo analysis and fuzzy sets. *Journal of Loss Prevention in the Process Industries, 44*, 503–514. doi:10.1016/j.jlp.2016.07.028

International Standards Organisation, Software engineering—Product quality Part 1: Quality model, SS-ISO/IEC 9126-1, 2003.

International Standards Organization. (2011). Systems and software engineering—systems and software quality requirements and evaluation (SQUARE)—systems and software quality models, ISO/IEC 25010.

International Telecommunication Union (UIT). (2005). *ITU Internet Reports*. The Internet of Things.

Ipema, A. H., Goense, D., Hogewerf, P. H., Houwers, H. W. J., & Van Roest, H. (2008). Pilot study to monitor body temperature of dairy cows with a rumen bolus. *Computers and Electronics in Agriculture, 64*(1), 49–52. doi:10.1016/j.compag.2008.05.009

Jain, A. K., & Dubes, R. C. (1988). *Algorithms for Clustering Data*. Englewood Cliffs, NJ: Prentice Hall.

Jan, B., Farman, H., Khan, M., Imran, M., Islam, I. U., Ahmad, A., ... Jeon, G. (2017). Deep learning in big data Analytics: A comparative study. *Computers & Electrical Engineering*. doi:10.1016/j.compeleccng.2017.12.009

Jang, R. (1992). *Neuro-Fuzzy Modelling: Architectures, Analysis and Applications* (PhD Thesis). University of California, Berkeley, CA.

Jaoua, A., & Mili, A. (1990). The Use of Executable Assertions for Error Detection and Damage Assessment. *J. Software Systems, 12*, 15-37.

Jarrow, R. A., Lando, D., & Turnbull, S. M. (1997). A Markov model for the term structure of credit risk spreads." -. *Review of Financial Studies*, *10*(2), 481–523. doi:10.1093/rfs/10.2.481

Jarrow, R. A., & Turnbull, S. M. (1995). Pricing derivatives on financial securities subject to credit risk. *The Journal of Finance*, *50*(1), 53–85. doi:10.1111/j.1540-6261.1995.tb05167.x

Jarrow, R. A., & Turnbull, S. M. (2000). The intersection of market and credit risk. *Journal of Banking & Finance*, *24*(1), 271–299. doi:10.1016/S0378-4266(99)00060-6

Jeffrey, D. B., & Raftery, A. E. (1993). Model-Based Gaussian and Non-Gaussian Clustering. *Biometrics*, *49*(3), 803–821. doi:10.2307/2532201

Jin, X., & Han, J. (2010). K-Medoids Clustering. In *Encyclopedia of Machine Learning* (pp. 564–565). Springer. doi:10.1007/978-0-387-30164-8_426

Juang, F. C., & Chin Lin, T. (1998). An On-Line Self Constructing Neural Fuzzy Inference Network and its applications. *IEEE Transactions on Fuzzy Systems*, *6*(1), 12–32. doi:10.1109/91.660805

Jung, K. M. (2017). liquidity risk and time-varying correlation between equity and currency returns. *Economic Inquiry*, *55*(2), 898–919. doi:10.1111/ecin.12418

Kamhoua, C. A., Pissinou, N., & Makki, S. K. (2010). Game Theoretic Analysis of Cooperation in Autonomous Multi Hop Networks. *Proc. of IEEE GLOBECOM*.

Kaplan, R. (2001). The nature of the view from home: Psychological benefits. *Environment and Behavior*, *33*(4), 507–542. doi:10.1177/00139160121973115

Kapoor, A., & Picard, R. W. (2002, May). Real-time, fully automatic upper facial feature tracking. In *Automatic Face and Gesture Recognition, 2002. Proceedings. Fifth IEEE International Conference on* (pp. 10-15). IEEE.

Kapoor, A., Burleson, W., & Picard, R. W. (2007). Automatic prediction of frustration. *International Journal of Human-Computer Studies, 65*(8), 724–736. doi:10.1016/j.ijhcs.2007.02.003

Kapoor, A., & Picard, R. W. (2005, November). Multimodal affect recognition in learning environments. In *Proceedings of the 13th annual ACM international conference on Multimedia* (pp. 677-682). ACM. 10.1145/1101149.1101300

Karami, A., Keiter, S., Hollert, H., & Courtenay, S. C. (2013). Fuzzy logic and adaptive neuro-fuzzy inference system for characterization of contaminant exposure through selected biomarkers in African catfish. *Environmental Science and Pollution Research International, 20*(3), 1586–1595. doi:10.100711356-012-1027-5 PMID:22752811

Karanki, D. R., Kushwaha, H. S., Verma, A. K., & Ajit, S. (2009). Uncertainty analysis based on probability bounds (P-box) approach in probabilistic safety assessment. *Risk Analysis, 29*(5), 662–675. doi:10.1111/j.1539-6924.2009.01221.x PMID:19302279

Kasabov, N., & Song, Q. E. (1999). *Dynamic Evolving Fuzzy Neural Networks with 'm-out-of-n' Activation Nodes for On-Line daptive Systems.* Technical Report TR99/04, Departement of Information Science, University of Otago.

Kaufman, L., & Rousseeuw, P. J. (1987). *Clustering by Means of Medoids in Statistical Data Analysis Based on the L_1 - Norm and Related Methods.* North- Holland.

Kawsar, F., Fitton, D., & Sundramoorthy, V. (2010). Smart Objects as Building Blocks for the Internet of Things. *IEEE Internet Computing, 14*(1), 44–51. doi:10.1109/MIC.2009.143

Kekäläinen, J., & Järvelin, K. (2002). Using graded relevance assessments in IR evaluation. *Journal of the Association for Information Science and Technology, 53*(13), 1120–1129.

Kelly, D. (2009). Methods for evaluating interactive information retrieval systems with users. *Foundations and Trends® in Information Retrieval, 3*(1–2), 1-224.

Kentel, E., & Aral, M. M. (2004). Probabilistic-Fuzzy Health Risk Modeling. *Stochastic Environmental Research and Risk Assessment, 18*(5), 324–338. doi:10.100700477-004-0187-3

Keßler, C. (2010). *Context-aware semantics-based information retrieval.* IOS Press.

Ketankumar, D. C., Verma, G., & Chandrasekaran, K. (2015). A Green Mechanism Design Approach to Automate Resource Procurement in Cloud. *Procedia Computer Science, 54*, 108 – 117. Doi:10.1016/j.procs.2015.06.013

Khandaker, M. (2009). Designing affective video games to support the social-emotional development of teenagers with autism spectrum disorders. *Annual Review of Cybertherapy and Telemedicine, 7,* 37–39. PMID:19592726

Klein, S. (2017). The World of Big Data and IoT. In *IoT Solutions in Microsoft's Azure IoT Suite* (pp. 3–13). Apress. doi:10.1007/978-1-4842-2143-3_1

Knight, W. (2016). AI Winter Isn't Coming. *MIT Technology Review*. Retrieved from https://www.technologyreview.com/s/603062/ai-winter-isnt-coming/

Kohonen, T. (1982). Self-organized formation of topologically correct feature maps. *Biological Cybernetics, 43*(1), 59–69. doi:10.1007/BF00337288

Kohonen, T. (1995). *Self-Organizing Maps* (Vol. 30). Berlin: Springer. doi:10.1007/978-3-642-97610-0

Kosko, B. (1992). *A Dynamical System Approach to Machine Intelligence. In Neural Networks and Fuzzy Systems*. Englewood Cliffs, NJ: Prentice Hall.

Kotwai, P. (n.d.). Green computing. *Scribd*. Retrieved from: https://www.scribd.com/document/140778684/51161349-Green-Computing

Koyama, K., Koyama, T., Matsui, Y., Sugimoto, M., Kusakari, N., Osaka, I., & Nagano, M. (2017). Characteristics of dairy cows with a pronounced reduction in first milk yield after estrus onset. *The Japanese Journal of Veterinary Research, 65*(2), 55–63.

Kringelbach, M. L., & Berridge, K. C. (2017). The affective core of emotion: Linking pleasure, subjective well-being, and optimal metastability in the brain. *Emotion Review*. PMID:28943891

Krishnapuram, R., & Keller, J. (1993). A Possibilistic Approach to Clustering. *IEEE Transactions on Fuzzy Systems, 1*(2), 98–110. doi:10.1109/91.227387

Kruijff, E., Marquardt, A., Trepkowski, C., Schild, J., & Hinkenjann, A. (2017). Designed emotions: Challenges and potential methodologies for improving multisensory cues to enhance user engagement in immersive systems. *The Visual Computer, 33*(4), 471–488. doi:10.100700371-016-1294-0

Kumar, R., & Kaur, G. (2011). Comparing Complexity in Accordance with Object Oriented Metrics. *International Journal of Computer Applications, 5*(8).

Kumar, H. (2017). Some Recent Defuzzification Methods. In *Theoretical and Practical Advancements for Fuzzy System Integration* (pp. 31–48). IGI-Global. doi:10.4018/978-1-5225-1848-8.ch002

Kumari, B., Kumar, V., Sinha, A. K., Ahsan, J., Ghosh, A. K., Wang, H., & DeBoeck, G. (2017). Toxicology of arsenic in fish and aquatic systems. *Environmental Chemistry Letters*, *15*(1), 43–64. doi:10.100710311-016-0588-9

Kumar, S., & Ng, B. (2001). Crowding and violence on psychiatric wards: Explanatory models. *Canadian Journal of Psychiatry*, *46*(5), 433–437. doi:10.1177/070674370104600509 PMID:11441783

Larson, K., & Picard, R. (2005, February). *The aesthetics of reading*. In Appears in Human-Computer Interaction Consortium Conference, Snow Mountain Ranch, Fraser, CO.

Le Ngo, A. C., See, J., & Phan, C. W. R. (2017). Sparsity in Dynamics of Spontaneous Subtle Emotion: Analysis & Application. *IEEE Transactions on Affective Computing*.

LeCun, Y., Bengio, Y., & Hinton, G. (2015). Deep learning. *Nature*, *521*(7553), 436–444. doi:10.1038/nature14539 PMID:26017442

Leung, Y., Zhang, J., & Xu, Z. (2000). Clustering by scale-space filtering. *IEEE Transactions on Pattern Analysis and Machine Intelligence*, *22*(12), 1396–1410. doi:10.1109/34.895974

Li, X., Lu, R., Liang, X., Shen, X., Chen, J., & Lin, X. (2011). Smart community: an internet of things application. IEEE Communications Magazine, 49(11), 68-75.

Limbourg, P., & de Rocquigny, E. (2010). Uncertainty analysis using evidence theory – confronting level-1 and level-2 approaches with data availability and computational constraints. *Reliability Engineering & System Safety*, *95*(5), 550–564. doi:10.1016/j.ress.2010.01.005

Lincke, R., Lundberg, J., & Löwe, W. (2008). Comparing software metrics tools. *Proceedings of the 2008 international symposium on Software testing and analysis*, 131–142.

Lin, T. C., & Lee, C. S. (1991). Neural Network Based Fuzzy Logic Control and Decision System. *IEEE Transactions on Computers*, *40*(12), 1320–1336. doi:10.1109/12.106218

Liu, W., Cui, Y., & Li, Y. (2015). Information systems security assessment based on system dynamics. *International Journal of Security and its Applications, 9*(2), 73-84.

Liu, J., Li, Y., Huang, G., Fu, H., Zhang, J., & Cheng, G. (2017). Identification of water quality management policy of watershed system with multiple uncertain interactions using a multi-level-factorial risk-inference-based possibilistic-probabilistic programming approach. *Environmental Science and Pollution Research International*, 1–21. doi:10.100711356-017-9106-2 PMID:28488149

Liu, J., Zhao, F., O'Reilly, J., Souarez, A., Manos, M., Liang, C.-J. M., & Terzis, A. (2010). *Project Genome: Wireless Sensor Network for Data Center Cooling*. Greenpeace International.

Liu, W., Wang, Z., Liu, X., Zeng, N., Liu, Y., & Alsaadi, F. E. (2017). A survey of deep neural network architectures and their applications. *Neurocomputing, 234*, 11–26. doi:10.1016/j.neucom.2016.12.038

Li, W. (1998). Another metric suite for object-oriented programming. *Journal of Systems and Software, 44*(2), 155–162. doi:10.1016/S0164-1212(98)10052-3

Li, W., & Henry, S. (1993). Maintenance metrics for the Object-Oriented paradigm. *Proceedings of First International Software Metrics Symposium*, 52-60. 10.1109/METRIC.1993.263801

Lloyd, S. (1982). Least Squares Quantization in PCM. *IEEE Transactions on Information Theory, 28*(2), 129–137. doi:10.1109/TIT.1982.1056489

Lokanath, S., Narayan, M. M., & Srikanta, P. (2016). Critical Heart Condition Analysis through Diagnostic Agent of e-Healthcare System using Spectral Domain Transform. *Indian Journal of Science and Technology, 9*(38). doi:10.17485/ijst/2016/v9i38/101937

Lokare, S. M. (2007). Commodity derivatives and price risk management: An empirical anecdote from India. *Reserve Bank of India Occasional Papers, 28*(2), 27-76.

Løvendahl, P., & Chagunda, M. G. G. (2010). On the use of physical activity monitoring for estrus detection in dairy cows. *Journal of Dairy Science, 93*(1), 249–259. doi:10.3168/jds.2008-1721 PMID:20059923

Lütkebohmert, E., Oeltz, D., & Xiao, Y. (2017). Endogenous credit spreads and optimal debt financing structure in the presence of liquidity risk. *European Financial Management, 23*(1), 55–86. doi:10.1111/eufm.12089

Lyu, M. R. (1996). Handbook of Software Reliability Engineering. IEEE Computer Society Press.

Lyu, M.R. (2007). Software Reliability Engineering: A Roadmap. *Future of Software Engineering*.

MacQueen, J. B. (1967). Some Methods for Classification and Analysis of Multivariate Observations. In *Proceedings of the 5th Berkeley Symposium on Mathematical Statistics and Probability*. Berkeley, CA: University of California Press.

Mader, T. L. (2003). Environmental stress in confined beef cattle. *Journal of Animal Science, 81*(14_suppl_2), E110-E119.

Mai, J. E. (2016). *Looking for information: A survey of research on information seeking, needs, and behavior*. Emerald Group Publishing.

Malviya, P., & Singh, S. (2013). A study about green computing. *International Journal of Advanced Research in Computer Science and Software Engineering, 3*(6), 790-794.

Marsit, I., Omri, M. N., & Mili, A. (2017). *Estimating the Survival Rate of Mutants*. Academic Press.

Martınez-Miranda, J., & Aldea, A. (2005). Emotions in human and artificial intelligence. *Computers in Human Behavior, 21*(2), 323–341. doi:10.1016/j.chb.2004.02.010

Martiskainen, P., Järvinen, M., Skön, J. P., Tiirikainen, J., Kolehmainen, M., & Mononen, J. (2009). Cow behaviour pattern recognition using a three-dimensional accelerometer and support vector machines. *Applied Animal Behaviour Science, 119*(1), 32–38. doi:10.1016/j.applanim.2009.03.005

Maxwell, R. M., Pelmulder, S. D., Tompson, A. F. B., & Kastenberg, W. E. (1998). On the development of a new methodology for groundwater-driven health risk assessment. *Water Resources Research, 34*(4), 833–847. doi:10.1029/97WR03605

McCall, J.A., Richards, P.K., & Walters, G.F. (1977). *Factors in software quality*. RADC TR-77-369, 1977. Vols I, II, III, US Rome Air Development center Reports NTIS AD/A-049 014, 015, 055.

McDuff, D., Gontarek, S., & Picard, R. (2014, August). Remote measurement of cognitive stress via heart rate variability. In *Engineering in Medicine and Biology Society (EMBC), 2014 36th Annual International Conference of the IEEE* (pp. 2957-2960). IEEE.

Melnykov, V., & Maitra, R. (2010). Finite Mixture Models and Model-Based Clustering. *Statistics Surveys, 4*(0), 80–116. doi:10.1214/09-SS053

Michalski, G. (2008). *Operational Risk in Current Assets Investment Decisions: Portfolio Management Approach in Accounts Receivable* [Agro Econ-Czech: Operační Risk v Rozhodování o Běžných Aktivech: Management Portfolia Pohledávek]. Academic Press.

Mili, A., & Tchier, F. (2015). Software Testing: Concepts and Operations. In Fundamentals of Software Engineering (2nd ed.). Academic Press.

Mili, A., Jaoua, A., Frias, A., & Helali, R. G. M. (2014). Semantic metrics for software products. *Innovations in Systems and Software Engineering, 10*(3), 203–217. doi:10.100711334-014-0233-3

Mills, E. (1998). *Software Metrics. SEI Curriculum Module SEI-CM-12-1.1*. Carnegie Mellon University, Software Engineering Institute.

Minsky, M. L. (1967). *Computation: finite and infinite machines*. Prentice-Hall, Inc.

Mitchell, R. S., Sherlock, R. A., & Smith, L. A. (1996). An investigation into the use of machine learning for determining oestrus in cows. *Computers and Electronics in Agriculture, 15*(3), 195–213. doi:10.1016/0168-1699(96)00016-6

Mitola, J., & Maguire, G. Q. (1999). Cognitive Radio: Making Software Radios More Personal. *IEEE Personal Communications, 6*(4), 13–18. doi:10.1109/98.788210

Mizzaro, S. (1997). Relevance: The whole history. *Journal of the Association for Information Science and Technology, 48*(9), 810–832.

Modares, M., Mullen, R., L., Muhanna, R., L. (2006). Natural Frequencies of a Structure with Bounded Uncertainty. *Journal of Engineering Mechanics*, 1363-1371.

Mofarrah, A., & Hussain, T. (2010). *Modeling for uncertainty assessment in human health risk quantification: A fuzzy based approach*. International congress on environmental modeling and soft modeling for environment's sake, fifth biennial meeting, Ottawa, Canada.

Mohanty, M. N. (2016). An Embedded Approach for Design of Cardio-Monitoring System. *Advanced Science Letters, 22*(2), 349–353. doi:10.1166/asl.2016.6854

Mohapatra, S. K., Palo, H. K., & Mohanty, M. N. (2017). Detection of Arrhythmia using Neural Network. *Annals of Computer Science and Information Systems, 14*, 97–100. doi:10.15439/2017KM42

Monma, H. (2011). FGCP/A5: PaaS Service Using Windows Azure Platform. *Fujitsu Scientific and Technical Journal, 47*(4), 443–450.

Morzy, T., Wojciechowski, M., & Zakrzewicz, M. (1999). Pattern-Oriented Hierarchical Clustering. *Proceeding of 3rd East Eur. Conf. Advances in Databases and Information Systems,* 179–190. 10.1007/3-540-48252-0_14

Moscadelli, M. (2004). *The modelling of operational risk: experience with the analysis of the data collected by the Basel Committee,* Academic Press.

Mottram, T., Lowe, J., McGowan, M., & Phillips, N. (2008). A wireless telemetric method of monitoring clinical acidosis in dairy cows. *Computers and Electronics in Agriculture, 64*(1), 45-48.

Nadiri, A. A., Gharekhani, M., Khatibi, R., & Moghaddam, A. A. (2017). Assessment of groundwater vulnerability using supervised committee to combine fuzzy logic models. *Environmental Science and Pollution Research International, 24*(9), 8562–8577. doi:10.100711356-017-8489-4 PMID:28194673

Nahar, J., Imam, T., Tickle, K. S., & Chen, Y. P. P. (2013). Association rule mining to detect factors which contribute to heart disease in males and females. *Expert Systems with Applications, 40*(4), 1086–1093. doi:10.1016/j.eswa.2012.08.028

Nasirahmadi, A., Edwards, S. A., & Sturm, B. (2017). Implementation of machine vision for detecting behaviour of cattle and pigs. *Livestock Science, 202,* 25–38. doi:10.1016/j.livsci.2017.05.014

Nauck, D. (1994). A Fuzzy Perceptron as a Generic Model for Neuro-Fuzzy Approaches. *Proc. Fuzzy-Systems, 2nd GI-Workshop.*

Nauck, D. (1995). Beyond Neuro-Fuzzy Systems: Perspectives and Directions. *Proc. of the Third European Congress on Intelligent Techniques and Soft Computing.*

Nauck, D., Klawon, F., & Kruse, R. (1997). *Foundations of Neuro-Fuzzy Systems.* J. Wiley & Sons.

Nauck, D., & Kurse, R. (1997). Neuro-FuzzySystems for Function Approximation. *4th International Workshop Fuzzy-Neuro Systems.*

Neslehova, J., Embrechts, P., & Chavez-Demoulin, V. (2006). Infinite mean models and the LDA for operational risk. *Journal of Operational Risk, 1*(1), 3–25. doi:10.21314/JOP.2006.001

Ning, H., & Wang, Z. (2011). Future Internet of Things Architecture: Like Mankind Neural System or Social Organization Framework. *IEEE Communications Letters, 15*(4), 461–463. doi:10.1109/LCOMM.2011.022411.110120

Okafor, K. C., Udeze, C. C., Ugwoke, F. N., Okoro, O. V., & Akura, M. (2013). Smart Green: A Novel Green Computing Architecture for Environmental Data Acquisition in IT/Automation Data Center. *African Journal of Computing and ICT, 6*(5).

Ordonez, C. (2006). Association rule discovery with the train and test approach for heart disease prediction. *IEEE Transactions on Information Technology in Biomedicine, 10*(2), 334–343. doi:10.1109/TITB.2006.864475 PMID:16617622

Oresko, J. J., Jin, Z., Cheng, J., Huang, S., Sun, Y., Duschl, H., & Cheng, A. C. (2010). A wearable smartphone-based platform for real-time cardiovascular disease detection via electrocardiogram processing. *IEEE Transactions on Information Technology in Biomedicine, 14*(3), 734–740. doi:10.1109/TITB.2010.2047865 PMID:20388600

Ortega, M., Pérez, M. Í., & Rojas, T. (2003). Construction of a Systemic Quality Model for Evaluating a Software Product. *Software Quality Journal, 11*(3), 219–242. doi:10.1023/A:1025166710988

Osowski, S., & Linh, T. H. (2001). ECG beat recognition using fuzzy hybrid neural network. *IEEE Transactions on Biomedical Engineering, 48*(11), 1265–1271. doi:10.1109/10.959322 PMID:11686625

Ozcift, A., & Gulten, A. (2011). Classifier ensemble construction with rotation forest to improve medical diagnosis performance of machine learning algorithms. *Computer Methods and Programs in Biomedicine, 104*(3), 443–451. doi:10.1016/j.cmpb.2011.03.018 PMID:21531475

Pacheco, M. A. C., & Vellasco, M. B. R. (2009). *Intelligent Systems in Oil Field Development under Uncertainty*. Berlin: Springer-Verlag. doi:10.1007/978-3-540-93000-6

Page, R. M. (1962). Man-Machine Coupling-2012 AD. *Proceedings of the IRE, 50*(5), 613–614. doi:10.1109/JRPROC.1962.288048

Pal, D., Mandana, K. M., Pal, S., Sarkar, D., & Chakraborty, C. (2012). Fuzzy expert system approach for coronary artery disease screening using clinical parameters. *Knowledge-Based Systems, 36*, 162–174. doi:10.1016/j.knosys.2012.06.013

Pal, N. R., Pal, K., & Bezdek, J. C. (1997). A Mixed c-Means Clustering Model. In *Proceedings of the 6th IEEE International Conference on Fuzzy Systems* (Vol. 1, p. 11). IEEE. 10.1109/FUZZY.1997.616338

Pal, N. R., Pal, K., Keller, J. M., & Bezdek, J. C. (2005). A Possibilistic Fuzzy c-Means Clustering Algorithm. *IEEE Transactions on Fuzzy Systems, 13*(4), 517–530. doi:10.1109/TFUZZ.2004.840099

Panjer, H. H. (2006). *Operational risk: modeling analytics* (Vol. 620). John Wiley & Sons. doi:10.1002/0470051310

Panksepp, J. (2004). *Affective neuroscience: The foundations of human and animal emotions*. Oxford University Press.

Panksepp, J. (2011). Cross-species affective neuroscience decoding of the primal affective experiences of humans and related animals. *PLoS One*, *6*(9), e21236. doi:10.1371/journal.pone.0021236 PMID:21915252

Panksepp, J. (2011). The basic emotional circuits of mammalian brains: Do animals have affective lives? *Neuroscience and Biobehavioral Reviews*, *35*(9), 1791–1804. doi:10.1016/j.neubiorev.2011.08.003 PMID:21872619

Parker, R. (2016). *Your Environment And You: Investigating Stress Triggers and Characteristics of the Built Environment*. Kansas State University.

Park, H. S., & Jun, C. H. (2009). A Simple and Fast Algorithm for K-Medoids Clustering. *Expert Systems with Applications*, *36*(2), 3336–3341. doi:10.1016/j.eswa.2008.01.039

Parkinson, H. (2015). Happy? Sad? Forget age, Microsoft can now guess your emotions. *The Guardian*. Retrieved from https://www.theguardian.com/technology/2015/nov/11/microsoft-guess-your-emotions-facial-recognition-software

Parsons, R., Tassinary, L. G., Ulrich, R. S., Hebl, M. R., & Grossman-Alexander, M. (1998). The view from the road: Implications for stress recovery and immunization. *Journal of Environmental Psychology*, *18*(2), 113–140. doi:10.1006/jevp.1998.0086

Pastoor, T. P., Bachman, A. N., Bell, D. R., Cohen, S. M., Dellarco, M., Dewhurst, I. C., ... Boobis, A. R. (2014). A 21st century roadmap for human health risk assessment. *Critical Reviews in Toxicology*, *44*(S3), 1–5. doi:10.3109/10408444.2014.931923 PMID:25070413

Pedronia, N., Zioa, E., Ferrariob, E., Pasanisid, A., & Couplet, M. (2012). Propagation of aleatory and epistemic uncertainties in the model for the design of a flood protection dike. PSAM 11 & ESREL, Helsinki: Finland.

Pedronia, N., Zioa, E., Ferrariob, E., Pasanisid, A., & Couplet, M. (2013). Hierarchical propagation of probabilistic and non-probabilistic uncertainty in the parameters of a risk model. *Computers & Structures*, *126*, 199–213. doi:10.1016/j.compstruc.2013.02.003

Picard, R. W. (1995). *Affective computing*. Academic Press.

Picard, R. W. (2001, June). Building HAL: Computers that sense, recognize, and respond to human emotion. In Photonics West 2001-Electronic Imaging (pp. 518-523). International Society for Optics and Photonics.

Picard, R. W., & Healey, J. (1997, October). Affective wearables. In *Wearable Computers, 1997. Digest of Papers., First International Symposium on* (pp. 90-97). IEEE.

Picard, R. W. (2002). Affective medicine: Technology with emotional intelligence. *Studies in Health Technology and Informatics*, 69–84. PMID:12026139

Picard, R. W. (2003). Affective computing: Challenges. *International Journal of Human-Computer Studies*, *59*(1), 55–64. doi:10.1016/S1071-5819(03)00052-1

Picard, R. W., & Du, C. (2002). Monitoring stress and heart health with a phone and wearable computer. *Motorola Offspring Journal*, *1*, 14–22.

Picard, R. W., Papert, S., Bender, W., Blumberg, B., Breazeal, C., Cavallo, D., ... Strohecker, C. (2004). Affective learning—a manifesto. *BT Technology Journal*, *22*(4), 253–269. doi:10.1023/B:BTTJ.0000047603.37042.33

Picard, R. W., & Picard, R. (1997). *Affective computing* (Vol. 252). Cambridge, MA: MIT Press.

Picard, R. W., & Rosalind, W. (2000). Toward agents that recognize emotion. *VIVEK-BOMBAY*, *13*(1), 3–13.

Picard, R. W., Vyzas, E., & Healey, J. (2001). Toward machine emotional intelligence: Analysis of affective physiological state. *IEEE Transactions on Pattern Analysis and Machine Intelligence*, *23*(10), 1175–1191. doi:10.1109/34.954607

Poh, M. Z., Kim, K., Goessling, A. D., Swenson, N. C., & Picard, R. W. (2009, September). Heartphones: Sensor earphones and mobile application for non-obtrusive health monitoring. In *Wearable Computers, 2009. ISWC'09. International Symposium on* (pp. 153-154). IEEE.

Poh, M. Z., McDuff, D. J., & Picard, R. W. (2011). Advancements in noncontact, multiparameter physiological measurements using a webcam. *IEEE Transactions on Biomedical Engineering*, *58*(1), 7–11. doi:10.1109/TBME.2010.2086456 PMID:20952328

Poh, M. Z., Swenson, N. C., & Picard, R. W. (2010). A wearable sensor for unobtrusive, long-term assessment of electrodermal activity. *IEEE Transactions on Biomedical Engineering*, *57*(5), 1243–1252. doi:10.1109/TBME.2009.2038487 PMID:20172811

Poria, S., Cambria, E., Bajpai, R., & Hussain, A. (2017). A review of affective computing: From unimodal analysis to multimodal fusion. *Information Fusion, 37*, 98–125. doi:10.1016/j.inffus.2017.02.003

Priovolos, T., & Duncan, R. C. (1991). *Commodity risk management and finance.* The World Bank.

Rabbachin, A., Quek, T. Q. S., & Shin, H. (2011). Cognitive Network Interference. *IEEE Journal on Selected Areas in Communications, 29*(2), 480–493. doi:10.1109/JSAC.2011.110219

Radjenović, D., Heričko, M., Torkar, R., & Živkovič, A. (2013). Software fault prediction metrics: A systematic literature review. *Information and Software Technology, 55*(8), 1397–1418. doi:10.1016/j.infsof.2013.02.009

Rahman, Q. A., Tereshchenko, L. G., Kongkatong, M., Abraham, T., Abraham, M. R., & Shatkay, H. (2015). Utilizing ECG-based heartbeat classification for hypertrophic cardiomyopathy identification. *IEEE Transactions on Nanobioscience, 14*(5), 505–512. doi:10.1109/TNB.2015.2426213 PMID:25915962

Ram, R., & Mohanty, M. N. (2017). The Use of Deep Learning in Speech Enhancement. *Annals of Computer Science and Information Systems, 14*, 107–111. doi:10.15439/2017KM40

Rębiasz, B., Bartłomiej, G., & Iwona, S. (2017). Joint Treatment of Imprecision and Randomness in the Appraisal of the Effectiveness and Risk of Investment Projects. *Information Systems Architecture and Technology: Proceedings of 37th International Conference on Information Systems Architecture and Technology–ISAT 2016–Part IV.*

Reynolds, C. J. (2005). *Adversarial uses of affective computing and ethical implications* (Doctoral dissertation). Massachusetts Institute of Technology.

Reynolds, C., & Picard, R. (2004, April). Affective sensors, privacy, and ethical contracts. In CHI'04 Extended Abstracts on Human Factors in Computing Systems (pp. 1103-1106). ACM. doi:10.1145/985921.985999

Riahi, A., Natalizio, E., Challal, Y., Mitton, N., & Iera, A. (n.d.). *A systemic and cognitive approach for IoT security.* Retrieved from https://hal.inria.fr/hal-00863955

Riseberg, J., Klein, J., Fernandez, R., & Picard, R. W. (1998, April). Frustrating the user on purpose: Using biosignals in a pilot study to detect the user's emotional state. In CHI 98 Cconference Summary on Human Factors in Computing Systems (pp. 227-228). ACM.

Rohrer, B. (2012, March). BECCA: Reintegrating AI for Natural World Interaction. *AAAI Spring Symposium: Designing Intelligent Robots*.

Rosenberg, L., Hammer, T., & Shaw, J. (1998). Software metrics and Reliability. *9th International Symposium on Software Reliability*.

Rosenblatt, F. (1958). The perceptron: A probabilistic model for information storage and organization in the brain. *Psychological Review, 65*(6), 386–408. doi:10.1037/h0042519 PMID:13602029

Rossi, B. (2015). *Gartner's Internet of Things predictions*. Information Age, Vitesse Media. Retrieved from http://www.information-age.com/technology/mobile-and-networking/123458905/gartners-internet-things-predictions

Russell, J., (2017, October). *A Dimensional Representation of Emotion*. Speech presented at ACII 2017, San Antonio, TX.

Russell, S., Norvig, P., & Intelligence, A. (1995). A modern approach. *Artificial Intelligence. Prentice-Hall. Egnlewood Cliffs, 25*, 27.

Ruthven, I. (2005). Integrating approaches to relevance. *New Directions in Cognitive Information Retrieval, 19*, 61-80.

Ruthven, I., Baillie, M., & Elsweiler, D. (2007). The relative effects of knowledge, interest and confidence in assessing relevance. *The Journal of Documentation, 63*(4), 482–504. doi:10.1108/00220410710758986

Rutten, C. J., Velthuis, A. G. J., Steeneveld, W., & Hogeveen, H. (2013). Invited review: Sensors to support health management on dairy farms. *Journal of Dairy Science, 96*(4), 1928–1952. doi:10.3168/jds.2012-6107 PMID:23462176

Saad, A., Fruehwirth, T., & Gervet, C. (2014). *The P-box CDF-Intervals: Reliable Constraint Reasoning with Quantifiable Information, to appear in Theory and Practice of Logic Programming*. TPLP.

Saini, N., Kharwar, S., & Agrawal, A. (2014). A Study of Significant Software Metrics. *International Journal of Engineering Inventions, 3*.

Sambuc, R. (1975). *Function Φ-Flous, Application a l'aide au Diagnostic en Pathologie Thyroidienne, These de Doctoraten Medicine*. University of Marseille.

Sano, A., & Picard, R. W. (2013, September). Stress recognition using wearable sensors and mobile phones. In *Affective Computing and Intelligent Interaction (ACII), 2013 Humaine Association Conference on* (pp. 671-676). IEEE.

Sano, A., Picard, R. W., & Stickgold, R. (2014). Quantitative analysis of wrist electrodermal activity during sleep. *International Journal of Psychophysiology, 94*(3), 382–389. doi:10.1016/j.ijpsycho.2014.09.011 PMID:25286449

Santillana Farakos, S. M., Frank, J. F., & Schaffner, D. W. (2013). Modeling the influence of temperature, water activity and water mobility on the persistence of Salmonella in low-moisture foods. *International Journal of Food Microbiology, 166*(2), 280–293. doi:10.1016/j.ijfoodmicro.2013.07.007 PMID:23973840

Santillana Farakos, S. M., Pouillot, R., Anderson, N., Johnson, R., Son, I., & Van Doren, J. (2016). Modeling the survival kinetics of Salmonella in tree nuts for use in risk assessment. *International Journal of Food Microbiology, 227*, 41–50. doi:10.1016/j.ijfoodmicro.2016.03.014 PMID:27062527

Santillana Farakos, S. M., Schaffner, D. W., & Frank, J. F. (2014). Predicting survival of Salmonella in low-water activity foods: An analysis of literature data. *Journal of Food Protection, 77*(9), 1448–1461. doi:10.4315/0362-028X.JFP-14-013 PMID:25198835

Saracevic, T. (2006). Relevance: A review of the literature and a framework for thinking on the notion in information science. Part II. In Advances in librarianship (pp. 3-71). Emerald Group Publishing Limited.

Saracevic, T. (1975). Relevance: A review of and a framework for the thinking on the notion in information science. *Journal of the Association for Information Science and Technology, 26*(6), 321–343.

Saracevic, T. (2007). Relevance: A review of the literature and a framework for thinking on the notion in information science. Part III: Behavior and effects of relevance. *Journal of the Association for Information Science and Technology, 58*(13), 2126–2144.

Sarangi, L. N., Mohanty, M. N., & Patnaik, S. (2016). Cardiac Diagnosis and Monitoring System Design using Fuzzy Inference System. *Rice (New York, N.Y.)*.

Sarangi, L., Mohanty, M. N., & Patnaik, S. (2016, June). Design of ANFIS Based E-Health Care System for Cardio Vascular Disease Detection. In *International Conference on Intelligent and Interactive Systems and Applications* (pp. 445-453). Springer.

Sarangi, L., Mohanty, M. N., & Patnaik, S. (2017). Detection of abnormal cardiac condition using fuzzy inference system. *International Journal of Automation and Control, 11*(4), 372–383. doi:10.1504/IJAAC.2017.087045

Sarangi, L., Mohanty, M. N., & Pattanayak, S. (2016). Design of MLP Based Model for Analysis of Patient Suffering from Influenza. *Procedia Computer Science*, *92*, 396–403. doi:10.1016/j.procs.2016.07.396

Scheirer, J., Fernandez, R., & Picard, R. W. (1999, May). Expression glasses: a wearable device for facial expression recognition. In CHI'99 Extended Abstracts on Human Factors in Computing Systems (pp. 262-263). ACM.

Schuller, B. W. (2017). IEEE Transactions on Affective ComputingźChallenges and Chances. *IEEE Transactions on Affective Computing*, *8*(1), 1–2. doi:10.1109/TAFFC.2017.2662858

Schutz, M. M., & Bewley, J. M. (2009). Implications of changes in core body temperature. In *Tri-State Dairy Nutrition Conference* (pp. 39-50). Academic Press.

Selim & Ismail. (1984). K-means-Type Algorithms: A Generalized Convergence Theorem and Characterization of Local Optimality. *IEEE Transactions on Pattern Analysis and Machine Learning, 6*(1), 81-87.

Senthil, S., & Senniappan, P. (2011). A Detailed Study on Energy Efficient Techniques for Mobile Ad hoc Networks. *International Journal of Computer Science Issues*, *8*(1), 383–387.

Shannon, C. E. (1953). Computers and automata. *Proceedings of the IRE*, *41*(10), 1234–1241. doi:10.1109/JRPROC.1953.274273

Sharkey, A., & Sharkey, N. (2011). Children, the elderly, and interactive robots. *IEEE Robotics & Automation Magazine*, *18*(1), 32–38. doi:10.1109/MRA.2010.940151

Sharma, S., & Sharma, G. (2016). Impact of Energy-Efficient and Eco-Friendly Green Computing. *International Journal of Computer Applications, 143*(7), 20-28. doi:10.5120/ijca2016910250

Shead, S. (2017). Facebook's AI boss: 'In terms of general intelligence, we're not even close to a rat'. *Business Insider*. Retrieved from http://www.businessinsider.com/facebooks-ai-boss-in-terms-of-general-intelligence-were-not-even-close-to-a-rat-2017-10

Shibly, A. (2015). *Green computing: - Emerging issue in IT*. Retrieved from: https://www.researchgate.net/publication/276266707_Green_Computing_-_Emerging_Issue_in_IT

Shults, M. C., Rhodes, R. K., Updike, S. J., Gilligan, B. J., & Reining, W. N. (1994). A telemetry-instrumentation system for monitoring multiple subcutaneously implanted glucose sensors. *IEEE Transactions on Biomedical Engineering, 41*(10), 937–942. doi:10.1109/10.324525 PMID:7959800

Sikchi, S. S., Sikchi, S., & Ali, M. S. (2013). Design of fuzzy expert system for diagnosis of cardiac diseases. *IJMSPH*. Retrieved from: http://citeseerx.ist.psu.edu/viewdoc/download?doi=10.1.1.301.1840&rep=rep1&type=pdf

Silverman, B. G., Hanrahan, N., Bharathy, G., Gordon, K., & Johnson, D. (2015). A systems approach to healthcare: Agent-based modeling, community mental health, and population well-being. *Artificial Intelligence in Medicine, 63*(2), 61–71. doi:10.1016/j.artmed.2014.08.006 PMID:25801593

Singh, M., Mohanty, M. N., & Choudhury, R. D. (n.d.). An Embedded Design for Patient Monitoring and Telemedicine. *International Journal of Electrical, Electronics, and Computer Engineering, 2*(2), 66-71.

Sloman, A. (1999). Review of affective computing. *AI Magazine, 20*(1), 127.

Snášel, V., Abraham, A., Owais, S., Platoš, J., & Krömer, P. (2010). User profiles modeling in information retrieval systems. In Emergent Web Intelligence: Advanced Information Retrieval (pp. 169-198). Springer London.

Soleymani, M., Garcia, D., Jou, B., Schuller, B., Chang, S. F., & Pantic, M. (2017). A survey of multimodal sentiment analysis. *Image and Vision Computing, 65*, 3–14. doi:10.1016/j.imavis.2017.08.003

Sperna Weiland, R. C., Laeven, R. J. A., & De Jong, F. (2017). Feedback between credit and liquidity risk in the us corporate bond market. *30th Australasian Finance and Banking Conference 2017.*

Spink, A., & Cole, C. (2005). A multitasking framework for cognitive information retrieval. *New directions in cognitive information retrieval*, 99-112.

Srivastava, V., & Motani, M. (2005). Cross-Layer Design: A Survey and the Road Ahead. *IEEE Communications Magazine, 43*(12), 112–119. doi:10.1109/MCOM.2005.1561928

Stein, C., Etzkorn, L., Cox, G., Farrington, Ph., Gholston, S., Utley, D., & Fortune, J. (2004). A New Suite of Metrics for Object-Oriented Software. *Proceedings of the 1st International Workshop on Software Audits and Metrics*, 49-58.

Steinhaus, H. (1957). Sur la Division Des Corps Matériels en Parties. *Bull. Acad. Polon. Sci., 4*(12), 801–804.

Stober, S. (2017). Toward Studying Music Cognition with Information Retrieval Techniques: Lessons Learned from the OpenMIIR Initiative. *Frontiers in Psychology*, *8*, 1255. doi:10.3389/fpsyg.2017.01255 PMID:28824478

Strauss, M., Reynolds, C., Hughes, S., Park, K., McDarby, G., & Picard, R. (2005). The handwave bluetooth skin conductance sensor. *Affective computing and intelligent interaction*, 699-706.

Su, C. J. (2008). Mobile multi-agent based, distributed information platform (MADIP) for wide-area e-health monitoring. *Computers in Industry*, *59*(1), 55–68. doi:10.1016/j.compind.2007.06.001

Sulzberger, S., Nschichold, E., & Vestli, S. (1993). FUN: Optimization of Fuzzy Rule Based Systems Using Neural Networks. *Proceedings of IEEE Conference on Neural Networks*, 312-316.

Sundmaeker, H., Guillemin, P., Friess, P., & Woelfflé, S. (2010). Vision and Challenges for Realising the Internet of Things. *CERPIoT- Cluster of European Research Projects on the Internet of Things.*

Tano, S., Oyama, T., & Arnould, T. (1996). Deep Combination of Fuzzy Inference and Neural Network in Fuzzy Inference. *Fuzzy Sets and Systems*, *82*(2), 151–160. doi:10.1016/0165-0114(95)00251-0

Tao, J., & Tan, T. (2005, October). Affective computing: A review. In *International Conference on Affective computing and intelligent interaction* (pp. 981-995). Springer. 10.1007/11573548_125

Thomas, L. C. (2000). A survey of credit and behavioural scoring: Forecasting financial risk of lending to consumers. *International Journal of Forecasting*, *16*(2), 149–172. doi:10.1016/S0169-2070(00)00034-0

Thomas, R. W., Friend, D. H., DaSilva, L. A., & Mackenzie, A. (2006). Cognitive Networks: Adaptation and Learning to Achieve End-to-End Performance Objectives. *IEEE Communications Magazine*, *44*(12), 51–57. doi:10.1109/MCOM.2006.273099

Tiwari, N. (2015). Green Computing. *International Journal of Innovative Computer Science & Engineering, 2*(1).

Tombros, A., Ruthven, I., & Jose, J. M. (2005). How users assess web pages for information seeking. *Journal of the Association for Information Science and Technology*, *56*(4), 327–344.

Toms, E. G., Freund, L., & Li, C. (2004). WiIRE: The Web interactive information retrieval experimentation system prototype. *Information Processing & Management, 40*(4), 655–675. doi:10.1016/j.ipm.2003.08.006

Tong, X., & Ban, X. (2014). A hierarchical information system risk evaluation method based on asset dependence chain. *International Journal of Security and its Applications, 8*(6), 81-88.

Tucker, W. T., & Ferson, S. (2003). *Probability bounds analysis in environmental risk assessment*. Applied Biomathematics.

Tupe, S., & Kulkarni, N. P. (2015). *ECA: Evolutionary Computing Algorithm for E-Healthcare Information System*. Retrieved from SPPU, Pune iPGCON-2015: http://spvryan. org/splissue/ipgcon/063. pdf

Uddin, M., & Rahman, A.A. (2012). Energy efficiency and low carbon enabler green IT framework for data centers considering green metrics. *Renewable and Sustainable Energy Reviews, 16*, 4078–4094. doi:10.1016/j.rser.2012.03.014

Ulrich, R. S., Simons, R. F., Losito, B. D., Fiorito, E., Miles, M. A., & Zelson, M. (1991). Stress recovery during exposure to natural and urban environments. *Journal of Environmental Psychology, 11*(3), 201–230. doi:10.1016/S0272-4944(05)80184-7

Urgaonkar, R., & Neely, M. J. (2008). *Opportunistic Scheduling with Reliability Guarantees in Cognitive Radio Networks. Proc. of IEEE INFOCOM*.

Vakkari, P., & Sormunen, E. (2004). The influence of relevance levels on the effectiveness of interactive information retrieval. *Journal of the Association for Information Science and Technology, 55*(11), 963–969.

Valido-Cabrera, E. (2006). *Software reliability methods*. Technical University of Madrid.

Vassalou, M., & Xing, Y. (2004). Default risk in equity returns. *The Journal of Finance, 59*(2), 831–868. doi:10.1111/j.1540-6261.2004.00650.x

Vipin Kumar, K. S., Albin, A. I., Arjunal, B., & Sheena, M. (2011). Compiler Based Approach To Estimate Software Reliability Using Metrics. *Annual IEEE India Conference*.

Vose, D. (2000). *Risk Analysis-A Quantitative Guide* (2nd ed.). Chichester, UK: John Wiley and Sons Ltd.

Vyzas, E., & Picard, R. W. (1999, May). Offline and online recognition of emotion expression from physiological data. In *Workshop on Emotion-Based Agent Architectures, Third International Conf. on Autonomous Agents,(Seattle, WA)* (pp. 135-142). Academic Press.

Waldren, S. E., Agresta, T., & Wilkes, T. (2017). Technology Tools and Trends for Better Patient Care: Beyond the EHR. *Family Practice Management, 24*(5), 28–32. PMID:28925624

Walker, K. A., Trites, A. W., Haulena, M., & Weary, D. M. (2011). A review of the effects of different marking and tagging techniques on marine mammals. *Wildlife Research, 39*(1), 15–30. doi:10.1071/WR10177

Wang, Z., Ji, M., & Sadjadpour, H. R. (2009). *Cooperation-Multiuser Diversity Tradeoff in Wireless Cellular Networks. Proc. of IEEE GLOBECOM.*

Ward, R. D., & Marsden, P. H. (2004). Affective computing: Problems, reactions and intentions. *Interacting with Computers, 16*(4), 707–713. doi:10.1016/j.intcom.2004.06.002

Weyuker, E. J. (1988). Evaluating Software Complexity Measures. *IEEE Transactions on Software Engineering, 14*(9), 1357–1365. doi:10.1109/32.6178

Wikipedia – Hardware Virtualization. (n.d.). Hardware virtualization. *Wikipedia.* Retrieved from: https://en.wikipedia.org/wiki/Hardware_virtualization

Wikipedia – Network Virtualization. (n.d.). *Network Virtualization.* Retrieved form https://en.wikipedia.org/wiki/network_virtualization

Wikipedia. (n.d.). Green computing. *Wikipedia.* Retrieved from: http://en.wikipedia.org/wiki/Green_computing

Wong, W.E., Debroy, V., & Restrepo, A. (2009). *The Role of Software in Recent Catastrophic Accidents.* IEEE Reliability Society 2009 Annual Technology Report.

Wren, C. R., & Reynolds, C. J. (2004). Minimalism in ubiquitous interface design. *Personal and Ubiquitous Computing, 8*(5), 370–373. doi:10.100700779-004-0299-2

Wright, A. (2011). Hacking cars. *Communications of the ACM, 54*(11), 18–19. doi:10.1145/2018396.2018403

Wruck, K.H., & Wu, Y.L. (2017). *Equity incentives, disclosure quality, and stock liquidity risk.* Academic Press.

Wu, M., Lu, T., Ling, F., Sun, J., & Du, H. (2010). Research on the architecture of Internet of things. *Proceedings of the 3rd International Conference on Advanced Computer Theory and Engineering (ICACTE'10)*, 484-487.

Xu, R., & Wunsch, D. II. (2005). Survey of Clustering Algorithms. *IEEE Transactions on Neural Networks, 16*(, 3), 645–677. doi:10.1109/TNN.2005.845141 PMID:15940994

Yan, H., Zheng, J., Jiang, Y., Peng, C., & Xiao, S. (2008). Selecting critical clinical features for heart diseases diagnosis with a real-coded genetic algorithm. *Applied Soft Computing, 8*(2), 1105–1111. doi:10.1016/j.asoc.2007.05.017

Yao, J., Dash, M., Tan, S. T., & Liu, H. (2000). Entropy-Based Fuzzy Clustering and Fuzzy Modeling. *Fuzzy Sets and Systems, 113*(3), 381–388. doi:10.1016/S0165-0114(98)00038-4

Yates, H., Chamberlain, B., & Hsu, W. (2017). *Spatially Explicit Classification Model for Affective Computing in Built Environments*. Paper presented at the Design for Affective Intelligence Workshop at ACII2017, San Antonio, TX.

Yates, H., Chamberlain, B., Norman, G., & Hsu, W. H. (2017, September). Arousal Detection for Biometric Data in Built Environments using Machine Learning. In *IJCAI 2017 Workshop on Artificial Intelligence in Affective Computing* (pp. 58-72).

Yoshino, M., Kubo, H., & Shinkuma, R. (2009). Modeling User Cooperation Problem in Mobile Overlay Multicast as a Multiagent System. *Proc. of IEEE GLOBECOM.*

Zadeh, L. A. (1965). Fuzzy Sets. *Information and Control, 8*(3), 338–353. doi:10.1016/S0019-9958(65)90241-X

Zahn, C. T. (1971). Graph-Theoretical Methods for Detecting and Describing Gestalt Clusters. *IEEE Transactions on Computers, 20*(1), 68–86. doi:10.1109/T-C.1971.223083

Zhang, X., & Shin, K. G. (2010). DAC: Distributed Asynchronous Cooperation for Wireless Relay Networks. *Proc. of IEEE INFOCOM.*

Zhang, H., & Zhu, L. (2011). Internet of Things: Key technology, Architecture and Challenging Problems. *Proceedings of IEEE International Conference on Computer Science and Automation Engineering (CSAE'11)*, 507-512.

Zhang, M., Wu, Q., Zheng, R., Wei, W., & Li, G. (2012). Research on Grade Optimization Self-Tuning Method for System Dependability Based on Autonomic Computing. *Journal of Computers, 7*(2), 333–340. doi:10.4304/jcp.7.2.333-340

Zhang, Q., Yang, L. T., Chen, Z., & Li, P. (2018). A survey on deep learning for big data. *Information Fusion, 42*, 146–157. doi:10.1016/j.inffus.2017.10.006

Zhang, X., Sun, M., Wang, N., Huo, Z., & Huang, G. (2016). Risk assessment of shallow groundwater contamination under irrigation and fertilization conditions. *Environmental Earth Sciences, 75*, 1–11.

Zheng, R., Wu, Q., Zhang, M., Li, G., Pu, J., & Wei, W. (2011). A self optimization mechanism of system service performance based on autonomic computing. *Journal of Computer Research and Development, 48*(9), 1676-1684. (in Chinese)

Zheng, R., Zhang, M., Wu, Q., Sun, S., & Pu, J. (2010). Analysis and application of bio-inspired multi-net security model. *International Journal of Information Security, 9*(1), 1–17. doi:10.100710207-009-0091-4

Zwietering, M. H. (2015). Risk assessment and risk management for safe foods: Assessment needs inclusion of variability and uncertainty, management needs discrete decisions. *International Journal of Food Microbiology, 213*, 118–123. doi:10.1016/j.ijfoodmicro.2015.03.032 PMID:25890788

About the Contributors

Pradeep Kumar Mallick is a Professor of Computer Science and Engineering at Vignana Bharati Institute of Technology, Hyderabad, India. His research interest is in Machine Intelligence, Image processing and Data Mining, and he has published more than 40 technical papers in the international journals and magazines of repute. In this credit published two Books that are Programming in C by laxmi Publication New Delhi, India and Research Advances in the Integration of Big Data and Smart Computing by IGI Global USA. He is a Gold Medalist in his Master. He is the Editor-in-Chief of the International Journal on Advanced Computer Theory and Engineering (IJACTE) and ITSI Transactions on Electrical and Electronics Engineering, as well as Chairman and Chief Mentor Institute for Research and Development India which is a professional body to encourage researchers and scholars to enhance their research throughput.

* * *

Dalila Amara is currently a Ph.D. student under the supervision of Prof. Latifa Rabai at Higher Institute of Management of Tunis, University of Tunis. Her research focuses on studying syntactic and semantic metrics use for software quality measurement.

Latifa Ben Arfa Rabai is a University Assistant professor in the Department of Computer Science at the Tunis University in the Higher Institute of Management (ISG). She received the computer science Engineering diploma in 1989 from the sciences faculty of Tunis and the PhD, from the sciences faculty of Tunis in 1992. Her research interest includes software engineering trends quantification, quality assessment in education and e-learning, and security measurement and quantification. She has published in information sciences Journal, IEEE Technology and Engineering Education magazine. She has participated in several international conferences covering topics related to the computer science, E-learning, quality assessment in education, cyber security.

Sukant Kishoro Bisoy received the Bachelor's degree in Computer Science & Engineering from (IE(India) and Master's degree in Computer Science & Engineering from Visvesvaraya Technological University (VTU), Belgaum, India in 2000 and 2003, respectively, and pursuing PhD degree in Computer Engineering from the SOA University, Bhubaneswar, India. He is an Assistant Professor of Computer Science & Engineering Department, C. V. Raman College of Engineering, Bhubaneswar, India. His current research interests are on 'Ad Hoc Network' and Wireless Sensor Network. He has served as Editor of Lecture notes on "Wireless Sensor networks: The recent Challenges". In addition, he has been involved in organizing many conferences, workshop and FDP. He has several publications in national and International conference and journals and given invited talk in many workshops.

Palash Dutta received his M. Sc. Degree in Mathematics from Dibrugarh University, Dibrugarh, India, in 2006, and his M. Phil. and Ph. D. degree in Mathematics from the same University in 2009 and 2012 respectively. He is an Assistant Professor in the Department of Mathematics, Dibrugarh University, since 2013. His research interests are uncertainty modelling using possibilistic approaches and decision making under the situations of uncertainty.

Sreedhar G. is working as a Associate Professor in the Department of Computer Science, Rashtiya Sanskrit Vidyapeetha (Deemed University), Tirupati, India since 2001. G. Sreedhar received his Ph.D in Computer Science and Technology from Sri Krishnadevaraya University, Anantapur, India in the year 2011. He has over 16 years of Experience in Teaching and Research in the field of Computer Science. He published more than 20 research papers related to web engineering in reputed international journals. He published 4 books from reputed international publications and he presented more than 20 research papers in various national and international conferences. He handled research projects in computer science funded by University Grants Commission, Government of India. He is a member in various professional bodies like academic council, board of studies and editorial board member in various international journals in the field of computer science, Information Technology and other related fields. He has proven knowledge in the field of Computer Science and allied research areas.

Parul Kalra is working as an Assistant Professor in the Department of Information Technology Amity School of Engineering & Technology, Amity University, Noida, Uttar Pradesh. She has 12 years of experience in the field of Academics and is actively involved in research & development activities. She is pursing Ph.D. (CSE) in the field of Cognitive Information Retrieval. She has completed her M.Sc. degree in Computer Science and M.Tech in Computer Science and Engineering from Banasthali Vidyapith, Rajasthan. Her area of interest includes Text Mining, Information Retrieval based on Cognitive and Personalized, Advance Database Management System. She has successfully published papers in national and international journals.

Harendra Kumar is an Assistant Professor in the Department of Mathematics & Statistics, Gurukula Kangari Vishwavidyalaya Haridwar-Uttarakhand (India).He obtained his Master of Science in 2003 from I.I.T Roorkee and Ph.D. from Gurukula Kangari Vishwavidyalaya in 2008. He has qualified NET (UGCJRF)- Dec.2003, NET(CSIR-JRF)-June 2005 and GATE-2004. He has published several research papers in National and International Journals in the field of Operation Research & Distributed Real Time System. He has more than 10 years teaching experience and has written three books for undergraduate & postgraduate students. He has authored several chapters for edited books. Dr. Kumar was invited to chair several international conference/workshop programs and sessions. He served in the program committees of several international conferences.

Deepti Mehrotra did Ph.D. from Lucknow University and currently she is working as Professor in Amity school of Engineering and Technology, Amity University, Noida, earlier she worked as Director of Amity School of Computer Science, Noida, India. She has more than 20 year of experience in research, teaching and content writing. She had published more than 100 papers in international refereed Journals and conference Proceedings. Her area of interest includes information retrieval and data mining, machine learning and knowledge engineering. She is currently working in domain of web usability, information retrieval and data mining

Brojo Kishore Mishra is an Associate Professor (IT) and the Institutional IQAC Coordinator at the C. V. Raman College of Engineering (Autonomous), Bhubaneswar, India. Received M.Tech and Ph.D. degrees in Computer Science from the Berhampur University in 2008 and 2012 respectively. Currently he is guiding 05nos. of Ph.D research scholar under Biju Pattnaik University of Technology, Odisha. His research interests include Data mining and big data analysis, machine learning, Soft computing, Evolutionary computation. He has already published more than 40 research papers in internationally reputed journals and referred conferences, 7 book chapters, has edited 1 book, and is acting as a member of the editorial board/ associate editor / Guest editor of various International journals. He served in the capacity of Keynote Speaker, Plenary Speaker, Program Chair, Proceeding chair, Publicity chair, Special session chairperson and as member of programme committees of many international conferences also. He was associated with a CSI funded research project as a Principal Investigator. He is a life member of ISTE, CSI, and member of IEEE, ACM, IAENG, UACEE, and ACCS.

Mihir Narayan Mohanty is presently working as a Professor in the Department of Electronics and Communication Engineering, Institute of Technical Education and Research. Siksha 'O' Anusandhan (Deemed to be University), Bhubaneswar, Odisha. He has published over 300 papers in International/National Journals and Conferences along with approximately 25 years of teaching experience. He is the active member of many professional societies like IEEE, IET, EMC & EMI Engineers India, ISCA, ACEEE, IAEng, CSI and also Fellow of IETE and IE (I). He has received his M.Tech. Degree in Communication System Engineering from the Sambalpur University, Sambalpur, Odisha and done his Ph.D. work in Applied Signal Processing. His area of research interests includes Applied Signal and image Processing, Intelligent Signal Processing, Biomedical Signal/Image Processing, Microwave Communication Engineering and Speech Processing.

Sunil Kumar Mohapatra received the B.Tech and M.Tech degree in Computer Science & Engineering from BPUT, Odisha .He was an former Assistant Professor in the department of Computer Science & Engineering, C. V. Raman College of Engineering, Bhubaneswar, India. Currently He is a faculty in the department of Computer Science & Engineering, College of Engineering Bhubaneswar. His current research interests are on 'Ad Hoc Network' and Wireless Sensor Network and Cloud Computing.

Index

A

accuracy 26, 247
activation function 17
affect 49-56, 58-66, 68
affective computing 49-56, 58-68
agriculture 50
animal sensors 49-52, 55-58, 62, 64-68
artificial intelligence 3-4, 59-61, 63-64, 68, 78-79

B

bell-shaped fuzzy numbers 101

C

cardiac disease 1, 9
CDSs 2
classification 1, 16-17, 26, 30, 67, 81, 86, 163, 203
cognitive computing 3-5
cognitive inference engine 35
Cognitive IoT 224, 227-228, 235, 238, 247, 254
COGNITIVE IOT INTERFERENCE 247
commodity price risk 183, 187

D

data center 129-131, 133, 136-138, 146
decision making 11, 236, 239-240
decision support 6, 30
Deep Neural Network 1, 16
Diabetes 5, 16-17, 26, 28, 30, 57

E

Eco-Friendly Computing 124
emotions 42, 45, 50, 52, 55, 59-61, 68
environmental sustainability 129-130

F

financial risk 183, 188, 190-193
foreign investment risk 183, 189, 191, 193
fuzzy numbers 99, 101-105, 107, 109-110, 113, 115, 117
fuzzy systems 78-82

G

Green IT 125, 146

H

health risk assessment 99-102, 104, 107-108, 113, 115, 117

I

inference engine 9, 35-38, 40, 43
information 1-10, 25-26, 29-30, 35-40, 42-43, 45, 53, 66, 100-103, 115, 133, 142, 150, 154, 159, 161, 163, 182, 193, 198-199, 225-229, 234-235, 239-240, 242, 244, 246, 248-250
information needs 35
information retrieval 35-40, 42-43, 198
Internet of Things 56, 224-227, 229-230, 235-236, 239, 247, 254

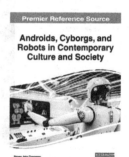

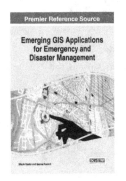

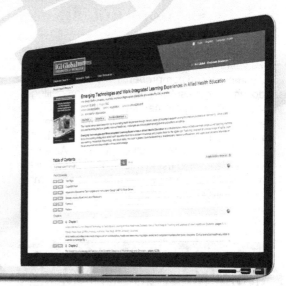

Printed in the United States
By Bookmasters